T0384469

ROBERT HOOKE

☞ Books in the RENAISSANCE LIVES series explore and illustrate the life histories and achievements of significant artists, rulers, intellectuals and scientists in the early modern world. They delve into literature, philosophy, the history of art, science and natural history and cover narratives of exploration, statecraft and technology.

Series Editor: François Quiviger

Already published

Albrecht Dürer: Art and Autobiography *David Ekserdjian*
Aldus Manutius: The Invention of the Publisher *Oren Margolis*
Artemisia Gentileschi and Feminism in Early Modern Europe *Mary D. Garrard*
Blaise Pascal: Miracles and Reason *Mary Ann Caws*
Botticelli: Artist and Designer *Ana Debenedetti*
Caravaggio and the Creation of Modernity *Troy Thomas*
Descartes: The Renewal of Philosophy *Steven Nadler*
Donatello and the Dawn of Renaissance Art *A. Victor Coonin*
Erasmus of Rotterdam: The Spirit of a Scholar *William Barker*
Filippino Lippi: An Abundance of Invention *Jonathan K. Nelson*
Giorgione's Ambiguity *Tom Nichols*
Hans Holbein: The Artist in a Changing World *Jeanne Nuechterlein*
Hieronymus Bosch: Visions and Nightmares *Nils Büttner*
Isaac Newton and Natural Philosophy *Niccolò Guicciardini*
Jan van Eyck within His Art *Alfred Acres*
John Donne: In the Shadow of Religion *Andrew Hadfield*
John Evelyn: A Life of Domesticity *John Dixon Hunt*
Leonardo da Vinci: Self, Art and Nature *François Quiviger*
Leon Battista Alberti: The Chameleon's Eye *Caspar Pearson*
Lucas Cranach: From German Myth to Reformation *Jennifer Nelson*
Machiavelli: From Radical to Reactionary *Robert Black*
Michelangelo and the Viewer in His Time *Bernadine Barnes*
Paracelsus: An Alchemical Life *Bruce T. Moran*
Petrarch: Everywhere a Wanderer *Christopher S. Celenza*
Piero della Francesca and the Invention of the Artist *Machtelt Brüggen Israëls*
Piero di Cosimo: Eccentricity and Delight *Sarah Blake McHam*
Pieter Bruegel and the Idea of Human Nature *Elizabeth Alice Honig*
Raphael and the Antique *Claudia La Malfa*
Rembrandt's Holland *Larry Silver*
Robert Hooke's Experimental Philosophy *Felicity Henderson*
Rubens's Spirit: From Ingenuity to Genius *Alexander Marr*
Salvator Rosa: Paint and Performance *Helen Langdon*
Thomas Nashe and Late Elizabethan Writing *Andrew Hadfield*
Titian's Touch: Art, Magic and Philosophy *Maria H. Loh*
Tycho Brahe and the Measure of the Heavens *John Robert Christianson*
Ulisse Aldrovandi: Naturalist and Collector *Peter Mason*

ROBERT HOOKE'S *Experimental Philosophy*

FELICITY HENDERSON

REAKTION BOOKS

For Paul, another inventive genius from the Isle of Wight

Published by Reaktion Books Ltd
Unit 32, Waterside
44–48 Wharf Road
London N1 7UX, UK
www.reaktionbooks.co.uk

First published 2024

Printed and bound in India by Replika Press Pvt. Ltd

A catalogue record for this book is available from the British Library

ISBN 978 1 78914 954 8

COVER: Robert Hooke's picture box, a portable camera obscura for tracing landscapes or coastlines, engraving from *Philosophical Experiments and Observations of the Late Eminent Dr Robert Hooke* (1726). Colouration by Katrin Idris (2004), photo akg-images.

CONTENTS

Introduction: Mad, Foolish and Phantastick

> I could proceed further, but methinks I can hardly forbear to blush, when I consider how the most part of Men will look upon this.[1]

obert Hooke is not known today for worrying about what people thought of him. Rescued from obscurity in the mid-twentieth century, and now regularly mentioned alongside his contemporaries Robert Boyle and Isaac Newton, he is generally remembered for two things: 'Hooke's Law', the principle governing the extension of springs; and his disagreement with Newton over gravitational theory. He has been characterized as a brilliant scientist, but a boastful and cantankerous colleague, a man who grew increasingly bitter over his contemporaries' failure to acknowledge his work, and who died in miserly squalor.

His peers may have partially recognized this figure, but they would have known it to be a caricature. His close friends would have thought back to the many evenings spent chatting in coffee-houses, the walks across London deep in conversation, afternoons drinking tea and playing chess, and shaken their heads. The greatest piece of evidence we have for Hooke's life comes from his own pen. His diary, or memoranda, as he called it, written in bursts over the course of 24 years, charts Hooke's progress from enthusiastic young man to respected mainstay of London's

1 Robert Hooke's picture box, a portable camera obscura for tracing landscapes or coastlines, engraving from *Philosophical Experiments and Observations of the Late Eminent Dr Robert Hooke* (1726).

scientific community. It shows that rather than being argumentative and self-absorbed, Hooke maintained long-standing friendships and close working relationships with men from a wide range of social spheres. More importantly, he considered these interactions to be crucial to his scientific method and to the development of science.

Hooke lived in exciting times. The English Civil Wars and Interregnum, the restoration of Charles II, the first coffee-houses and female actresses, the Great Plague of 1665–6, the Great Fire of London, the Glorious Revolution: Hooke witnessed all of them. These stirring events, though, were just a backdrop for what Hooke and his colleagues would have considered to be the truly exciting development of the period: the European scientific revolution. Sparked by the writings of Francis Bacon and stoked by Galileo's celestial observations, William Gilbert's treatise on magnetism and William Harvey's discovery of the circulation of blood, the reformation of knowledge was in full swing by 1660. In London, the Royal Society was founded to promote the study of the natural world through experiments and observation. Hooke was the society's first paid employee, and he remained one of its central figures for the rest of his life.

As a key fellow of the Royal Society, Hooke helped to shape the formation of a new scientific method suitable for the aims of London's first scientific institution. The fellows looked back to Francis Bacon as the great promoter of the new science at the turn of the seventeenth century, but Bacon's own method was far from complete. Hooke and his contemporaries needed to experiment not only with the phenomena of the natural world, but with their own scientific methodology. Many familiar aspects of modern science were yet to emerge at this time. There was no formal peer-review system, and questions about the competence and trustworthiness of practitioners were often hard to resolve. There was not yet a regular format, or language, for communicating scientific ideas. An accepted standard for what constituted scientific evidence was still evolving. Our modern disciplines of physics, biology and so on were just emerging, and it was common for people to contribute to discussions on many different topics. There was no assumption, as there sometimes is today, of science being inimical to religion: many of Hooke's colleagues were deeply religious and felt that their work would lead humankind closer to God. Ideas about scientific objectivity were emerging, but the modern ideal was not yet formulated. Even the word 'scientist' did not exist: Hooke and his colleagues were philosophers, usually distinguishing themselves as new or experimental philosophers because they prioritized experiment and observation as methods of understanding the natural world.

Uniquely among his contemporaries, Hooke wrote his own instructions for a method – or as he envisaged it, an 'engine' – that would support the mind in its philosophical enquiries. Because he saw algebra as a kind of device for solving problems in geometry, he called his method a 'Philosophical Algebra', or 'an Art of directing the Mind in the search after Philosophical Truths'. Just as regular algebra provided a structured way to work through mathematical problems, his philosophical algebra would make it 'very easy to proceed in any Natural Inquiry, regularly and certainly'. Hooke's own algebra is incomplete. It was to be in two parts: the first explained the 'manner of Preparing the Mind, and Furnishing it with fit Materials to work on'; the second would demonstrate the rules and methods for working with this initial data.[2] Only the first part exists now (or perhaps more likely, only the first part was ever written). Yet even this is an unusually systematic description of early modern scientific practice. None of his colleagues at the Royal Society attempted anything of comparable length or detail. It gives us a fascinating insight into Hooke's working methods and helps to explain some of the seemingly unrelated aspects of his working life: the lack of consistency in his subject-matter, his interest in mechanics and instruments and his practice of gathering information from conversations and books. Perhaps most importantly for a biographer, it focuses on how the experimental philosopher should behave. It is a blueprint for how an ordinary person could begin to think and act like a scientist. If we take it seriously, we can read here a description of who Hooke thought he was.

This focus on Hooke's own method is a new approach to his biography. Scholars in the twentieth century rediscovered Hooke, admiring in particular the breadth of his accomplishments. Historians of science have assessed his contributions to a wide array of fields, and the most recent book-length study by Francesco G. Sacco provides a rich and detailed analysis of Hooke's natural philosophy as a whole. There are also several excellent biographies of Hooke, most recently by Lisa Jardine and Stephen Inwood.[3] The present work attempts neither a full consideration of Hooke's research nor a detailed biography. Instead, I want to explore how Hooke's scientific methodology was reflected in his life and work, and by doing so, perhaps counter some of the enduring myths about Hooke and explain some of his apparent peculiarities. Faced with what seemed like a relentless churn of ideas and discoveries, historians have labelled Hooke as restless, almost too active, rarely finishing his projects and not publishing the kinds of weighty treatises that some of his contemporaries produced for posterity. Scholars have also queried Hooke's status in the scientific community of the time, doubting that he was gentlemanly enough to gain his colleagues' respect, and suggesting that he remained tainted by his role as paid employee at the Royal Society. His disputes with various colleagues and other philosophers seem to support this unsociable image. These are topics to which I will return in the following chapters, but it seems to me that these aspersions say more about what a modern audience wants from historical figures, or from the story of their lives, than about the man himself. By assessing Hooke in light of his own work, often expressed in his own words, I hope we might be able to see beyond our own preconceptions to understand what Hooke himself thought he was doing.

Hooke is a particularly intriguing figure because he was uniquely positioned to take advantage of London as a hub of mercantile activity, manufacturing and information exchange. Alongside his scientific work as 'curator of experiments' at the Royal Society, he also became one of the official surveyors of the City of London after the Great Fire in 1666, taking a leading role in the process of rebuilding the city. Restoration London was a small place by modern standards, and Hooke would have been well known; he certainly had a wide circle of acquaintances from different social spheres. As the epigraph at the head of this Introduction suggests, Hooke was painfully aware of his image as a virtuoso. The word originally meant a scholar, intellectual or philosopher, but in the period it became increasingly negative: someone who came up with wildly speculative ideas that his contemporaries thought were, in Hooke's own words, 'seemingly mad, foolish and phantastick'. The suspicion, incomprehension and downright derision that greeted the new philosophy and its practitioners is perhaps the most crucial difference between science in Hooke's day and our own. But Hooke's answer to these kinds of criticisms was simple: no one ever advanced human knowledge by assuming that certain things were impossible to know or to do. Believing that anything might be possible meant taking new ideas seriously: the first step towards turning those ideas into reality. At the heart of Hooke's philosophical method was the understanding that the new, experimental, philosophy – what we would today call science – had the potential to improve human life beyond recognition.

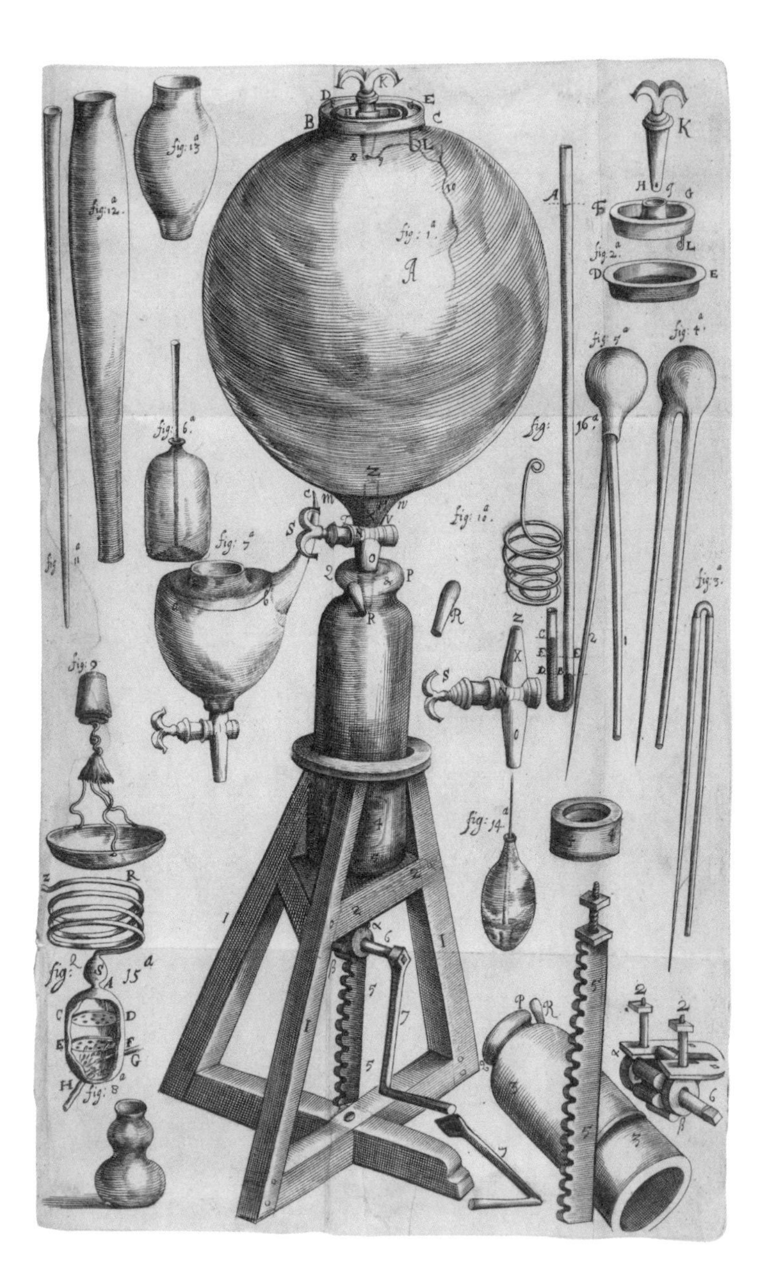

fig: 1a
fig: 2a
fig: 3a
fig: 4a
fig: 5a
fig: 6a
fig: 7a
fig: 8a
fig: 9
fig: 10a
fig: 11a
fig: 12a
fig: 13a
fig: 14a
fig: 15a
fig: 16a

ONE

The Present Deficiency of Natural Philosophy

The Business of Philosophy is to find out a perfect Knowledge of the Nature and Proprieties of Bodies, and of the Causes of Natural Productions, and this Knowledge is not barely acquir'd for itself, but in order to the inabling a Man to understand how . . . he may be able to produce and bring to pass such Effects, as may very much conduce to his well being in this World.[1]

hen Robert Hooke sat down to write these first lines of his scientific method he was a young man, but one who had already embarked on the quest that would occupy him for the rest of his life. His statement of intent for natural philosophy links knowledge with the power to change the world. A disciple of Francis Bacon, Hooke believed that the new knowledge he sought must be useful for improving people's lives: 'the reliefe of Mans estate', as Bacon had put it.[2] Hooke wrote these lines because he was asking himself why philosophy had not yet advanced very far, in spite of 'many Men, in divers Ages of the World' having tried to make progress.[3] The reason why so little progress had been made, Hooke suggested, was partly because earlier philosophers had never made use of a proper method. This was the problem Hooke set out to solve in his own writing. He agreed with Bacon that our own fallible human nature actually works against a true understanding of the

2 Robert Boyle's air pump, 1660. Specimens could be put inside the large glass globe at the top and the air pumped out, creating a partial vacuum.

natural world, but Hooke argued that there was a solution: the method for making experiments and observations, and recording and using the results, that he called his 'Philosophical Algebra'.

Remarkably few of Hooke's contemporaries attempted to systematize or write down a new method for making knowledge, despite a dedicated community of those interested in the experimental philosophy. Why did Hooke think he could answer this question? He seems to have been curious about the world around him since childhood. Born in 1635 at Freshwater, a small village on the Isle of Wight, he was the youngest of the four children of John Hooke, the curate of the parish church, and his second wife, Cecilie (née Gyles). Growing up, he tells us, he was sickly and his frequent headaches prevented much formal education; but he had an aptitude for mechanisms, and he recalled making a working wooden clock and a small wooden ship complete with rigging, masts and 'a contrivance to make it fire off some small Guns' as it sailed.[4] His father died in 1648, when Hooke was thirteen, and instead of staying on the island with his mother and elder siblings he took his small inheritance and travelled to London. His early biographers agree that he spent some time as apprentice to the portrait painter Sir Peter Lely, but that shortly afterwards he went to live with Dr Richard Busby, the headmaster of Westminster School. With Busby he learnt mathematics and music, and improved his Latin and Greek.

His experimental life began in Oxford, where he moved in 1653 or 1654 to attend university. In Oxford he met John Wilkins, Christopher Wren, Seth Ward, Thomas Willis and Robert Boyle, men who became mentors, fellow experimenters and friends. There was a thriving experimental community in Interregnum Oxford, which offered its members a chance to put aside the controversy and divisions of the Civil Wars and concentrate on astronomy, physics and anatomy. Willis was a successful physician who conducted chemical and anatomical experiments, later publishing an influential book depicting the anatomy of the brain. He employed Hooke as an assistant, and he must have been impressed with Hooke's dexterity because he recommended him to Robert Boyle. Boyle was only eight years older than Hooke, but as the youngest son of Richard Boyle, first earl of Cork, he was independently wealthy. He had moved to Oxford in the mid-1650s and began a concentrated programme of research and writing. Hooke helped Boyle with his experiments on air, many of which used a vacuum chamber known as the 'air pump' (illus. 2). This was an early test of Hooke's mechanical skills. He had seen an air pump made for Boyle by the instrument-maker Ralph Greatorex, which Hooke later recalled was 'too gross to perform any great matter'.[5] Hooke and Boyle built a more successful model, which Hooke operated. Expensive, delicate and temperamental, the air pump became a defining symbol of the new science, and Boyle's experiments with it strongly influenced Hooke's own science.

Hooke was Boyle's social inferior and employee, and he always retained a sense of duty and respect towards his early patron, but snippets of information suggest that the two men grew together as experimental philosophers, learning from each other. Boyle commented that it was Hooke who helped him to understand the philosophy of the influential French thinker René Descartes, and together they studied Euclid's *Elements*, the foundational mathematical textbook of the time. The two men must also have had many conversations about the usefulness of science, and practical discussions about scientific method.[6] Like Hooke, Boyle was strongly influenced by Bacon's writings. Although neither followed Bacon slavishly, both agreed with Bacon's overriding belief that current knowledge of the natural world was embarrassingly defective, and broadly agreed with his suggestions for improvement. Separately, both Hooke and Boyle elaborated on some of Bacon's key points: collecting and recording data would be key to the new science; preparing the mind for rational thought was an important first step; and collaboration would be necessary.

It was the drive towards a collaborative scientific endeavour that finally led Hooke to leave Boyle's employment. In 1660 a small group of men who had been meeting informally to discuss the new philosophy decided to incorporate themselves to become England's first scientific institution, the Royal Society of London for Improving Natural Knowledge (soon shortened informally to the Royal Society). Charles II's restoration to the throne that year had prompted some of Hooke's Oxford friends to move to London. They were prominent among the founding fellows: Boyle, John Wilkins, Seth Ward and Christopher Wren were present at the first meeting. They were joined by courtiers, physicians and other gentlemen interested in the new philosophy. Slowly the group expanded, but it became clear that the early fellows were enthusiastic but mostly either too busy or too inexperienced to drive an institutional research programme. What they needed was a 'curator of experiments' to do the hands-on work. In November 1662 the courtier and politician Sir Robert Moray announced that he had found 'a person willing to be employed as a curator by the society, and offering to furnish them every day, on which they met, with three or four considerable experiments'. Furthermore, Moray explained that this person did not expect to be paid for his services until the society was in a less precarious financial situation. The altruistic person was Robert Hooke. The proposal was approved unanimously, and Hooke was asked to bring to the weekly meetings both the 'three or four' experiments of his own, as well as 'such others, as should be mentioned to him by the society'.[7] His work with Boyle and Willis at Oxford had demonstrated his ability to design, construct and manipulate experimental apparatus, to reflect on the results of experiments and to design innovative trials in order to test a new train of thought. This skillset was highly unusual, and it promoted Hooke almost immediately to a key position within the Royal Society. Despite joining as a kind of unpaid assistant, his contributions to meetings were significant in terms of both the experiments that he demonstrated in front of his colleagues and the ensuing discussion. Six months later, in June 1663, the society's council signalled their approval by electing Hooke as a fellow and exempting him from the normal subscription fees.[8]

After this promising start, Hooke's contributions to Royal Society meetings continued in their frequency and significance over the next four decades. Very few of those early meetings went by without an experimental demonstration, observation or report from Hooke, or at the very least a comment added to the discussion. In his first year as curator of experiments he reported on, or was asked to consider, a daunting collection of research topics or questions: the properties of air and water; the crystalline structures of frozen urine, water and snow; gravity; combustion; how to make coloured glass; different kinds of fuel; questions about natural history to be sent to Greenland; the force and speed of falling bodies; his observations of a lunar eclipse; an account of a newly invented cart with legs instead of wheels; how to take samples of water from the bottom of the sea; his microscopical observations; an account of a Chinese cart with one wheel; his observations of the satellites of Jupiter; his observations of petrified wood; wheat sinking and floating in water; Prince Rupert's machine for raising water; instruments for sounding the sea; fish anatomy; an instrument to determine the strength of gunpowder; 'stones' found (post-mortem) in the heart of the earl of Balcarres; recommendations for a method of recording the weather; a hygroscope made from the beard of a wild oat; and an experiment in which a piece of a dog's skin was cut off and sewn on again. He was also put in charge of

the 'Repository', the Royal Society's collection of natural and man-made artefacts. He performed his role with dedication and enthusiasm, even though privately he may have had doubts about the significance of some of these topics: he seems not to have taken the cart with legs particularly seriously (illus. 3).

The Royal Society's institutional business centred on their weekly meetings. Here the fellows read and discussed correspondence and reports on experiments and observations, and shared news of philosophical interest. They watched experiments (often demonstrated by Hooke) and discussed the results, offering corroborating evidence or advice about further avenues to be explored. The minutes of the meetings demonstrate the value in a group of well-connected gentlemen, courtiers and aristocrats joining together to pursue all kinds of questions. Members with contacts in foreign parts were given lists of queries about local conditions in the East Indies, Africa and Scandinavia; visiting dignitaries and ambassadors from Italy, Denmark and elsewhere attended meetings and promised to

3 Hooke's drawing of the cart with legs designed by Francis Potter, 1663. In his assessment of the cart he pointed out that it would not reverse very easily.

assist with enquiries; landowners were asked for information about coalmines on their estates; and the Duke of Buckingham promised to bring in a piece of unicorn's horn.[9]

While some of the fellows pursued their own research agendas individually in their own time, the emphasis at Royal Society meetings was on sharing results and trying to establish 'matters of fact': observations about the natural world that could be verified, agreed upon and documented. Discussions ranged over almost every conceivable topic. The following page of minutes records a conversation in July 1678 sparked by the demonstration of what happened when an eel was put into the air pump and the air evacuated (unsurprisingly, the eel died). The conversation, even in this tersely minuted format, epitomizes the kind of wide-ranging, loosely connected exchange of facts and ideas that characterized the early meetings of the Royal Society:

> Upon discoursing about the food of fishes, Dr Croune related, that fishmongers never find anything in the maws of salmon. And that a lady, who had been very inquisitive in that kind, had observed the same: but that the contrary was observed in most other fish.
>
> Dr Grew remarked, that the guts in salmon placed round the stomach answering the *intestinum caecum* in other animals were very full, though the stomach [*sic*] were empty.
>
> Sir John Hoskyns related, that there was one sort of whale towards the north, that was reported to feed upon flies; vast quantities of which had been found in the stomachs of those whales: and that their fins seemed to serve for the straining of the flies from the water.
>
> Mr Hooke gave an account of the structure of the mouth and fins of that whale, which was cast on shore at Greenwich about twenty years before.

> He also mentioned a late relation, which he had seen printed in Low Dutch, of a voyage to Spitzbergen or Greenland, wherein was a description of that sort of whale, together with pictures of them.
>
> Mr Henshaw affirmed, that salmons feed upon flies.
>
> The bishop of Chester remarked, that he had a dish of fish, which had been taken very whole out of the maw of a large fish; that they eat very well; and that they were a sort of flounders.
>
> A discourse was then occasioned about poisons.[10]

Although the discussion began with a live experiment, the fellows quickly moved beyond the immediate fate of the eel to wider considerations of fish, their anatomy and their diets. They offered evidence from a variety of sources: Dr Croune cited fishmongers and an inquisitive lady; Nehemiah Grew, a pioneering comparative anatomist, commented on findings from his own dissections; Hooke reported his twenty-year-old recollection of a whale stranded at Greenwich and a more recent Dutch book; Sir John Hoskins related some information gathered by hearsay; and Thomas Henshaw and John Wilkins, now the bishop of Chester, reported from their own experience. Wilkins's revelation that he had eaten a dish of fish that had been found in the stomach of a larger fish may or may not have prompted the conversational turn towards poisons.

Hooke was almost always part of conversations such as these, no matter what topic was being discussed. Despite his own concerns about it, he had a retentive memory, and as we shall see in later chapters, he gathered information from a wide range of sources. For Hooke, in-depth investigations into a single topic such as light, the properties of air, comets or indeed what fish eat would form the foundation for the new philosophy. He described them as 'the Repository of Materials, out of which a new and sound Body of Philosophy may be raised'.[11] His choice of metaphors here is significant. In addition to his other responsibilities, he was the curator of the Royal Society's Repository, the somewhat random collection of man-made and natural items donated to, or bought by, the society and housed in Gresham College. These items ranged from human skeletons (male and female), through the animal kingdom, plant specimens and minerals, to man-made objects including scientific instruments, artefacts from foreign cultures, paintings and examples of Chinese and Arabic writing. Together they formed part of the set of jigsaw pieces of information about the world that Hooke and his colleagues were slowly collecting.

Though increasingly busy on the Royal Society's behalf, Hooke was still not earning a salary for his work. The society's meetings took place at Gresham College, an institution founded to provide free lectures for Londoners in fields such as law, theology and music (illus. 4). Christopher Wren was professor of astronomy at Gresham College. Hooke applied for the position of professor of geometry when it became vacant in 1664, but he was unsuccessful. Shortly afterwards, however, Sir John Cutler offered to pay Hooke £50 a year to give a series of lectures at Gresham College on a topic to be decided by the Royal Society. This was a generous offer: relative to an average earner it would work out as approximately £188,000 in today's terms.[12] Cutler was a wealthy merchant and prominent figure in the City of London, and it was appropriate that the topic of the Cutlerian lectures was designated as 'the History of Nature and Art'.[13] Art in this sense meant human workmanship. Cutler was elected to an honorary membership of the society in thanks for this donation to the cause of science, which the society hoped would be the first of many from wealthy benefactors (it was not). Soon after, in January 1665, the society elected Hooke as its 'curator'. This position, granted 'for perpetuity', made official

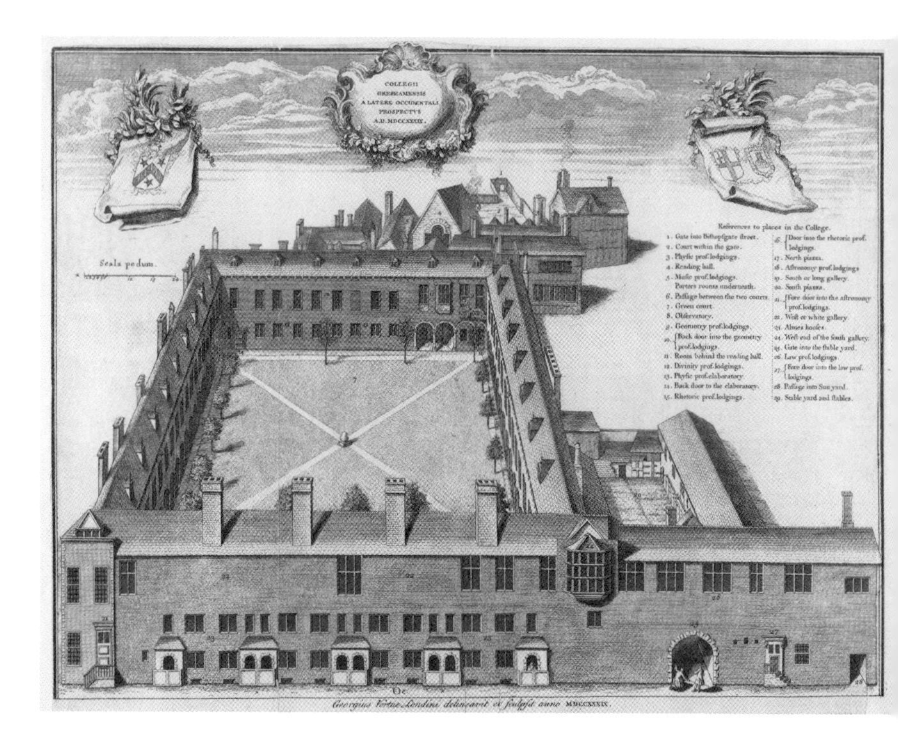

the work that Hooke had previously been doing as a volunteer. The society had initially decided on an annual salary of £80 for this post, but with Hooke's agreement they reduced it to £30 in light of Cutler's donation. As others have pointed out, these appointments effectually made Hooke the first professional English scientist. When the recent election of Arthur Dacres to the Gresham professorship of geometry was found to be flawed and was overturned, Hooke was elected in his place in March 1665. He was by this point already living at Gresham College, and his official post there reinforced his place at the heart of London's scientific community.

Hooke's early writings convey his excitement and optimism about the work he was doing. He published his longest and

4 A view of Gresham College, London, 1739. Hooke's rooms were in the rear right corner of the court; his rooftop observatory is visible.

best-known work, *Micrographia*, in January 1665: Samuel Pepys spotted the lavishly illustrated book of microscopic observations at his bookseller's on 2 January and immediately ordered a copy for himself. Hooke's fascination with the 'new visible World' he explored through his microscope is obvious to his readers, and infectious. Each discovery sparked more questions for him. How do sponges grow? Could the bubbling of natural springs be caused by the same power that made water rise in a narrow pipe? Do all insects originate from eggs? Hooke invited his readers to join him in his observations, patiently describing his microscope and working method so that those who could afford the expensive equipment might be able to replicate his findings. It was becoming clear to the scientific community that their contemporaries were more willing to laugh at them than to join their endeavours, but Hooke remained convinced that if only his readers could see what he had seen, they would understand that science had the potential to change their lives for the better.

Hooke's thinking about scientific method was possibly prompted, and certainly informed, by his microscopy. His project had involved producing a long series of microscopic observations for the fellows at their weekly meetings and then finding a way to communicate his findings to a general audience. This, alongside what we can presume was his dawning realization that very few of the fellows were fit to add much of substance to the society's scientific programme, must have prompted him to think more about how others could replicate his own success at experimental work. His preface to *Micrographia* sketched out some aspects of his practice, but his larger scheme, probably written shortly after *Micrographia* was published, went into much more detail. He began with a call for change: the current state of knowledge was defective, and old ways of doing philosophy had not improved things very much. The question many of his contemporaries would have asked was why anyone might want to bother doing experimental philosophy at all? What was the point? Hooke's answer was simple, stated with great certainty and optimism in the quote at the beginning of this chapter: by knowing more about the world, mankind would be able to shape nature to improve human life. Having dispatched that problem, Hooke went on to introduce his 'Philosophical Algebra'. Like algebra in geometry, Hooke believed that by following his method the mind would be led inexorably towards the truth. Although he referred to his algebra regularly, he seems never to have written the full method down. From the hints he gave, it seems likely to have been a system of organizing and recording information in such a way as to make it simpler to develop theories based on basic principles. However, the fact that he never explained it in detail suggests that it was not so much a part of his working practice as he might have wanted it to be. His existing method concentrates on the first part of the problem, which he described as the 'manner of Preparing the Mind, and Furnishing it with fit Materials to work on'.[14] In a sense, this was the more important stage of the process because it formed a foundation for any future work.

The first step in preparing the mind to work philosophically was to acknowledge that human nature is imperfect; the mind and the body are not designed for the tasks of experimenting and observing. For Hooke, solving this problem relied on two things: using scientific instruments to regulate and extend the senses; and being constantly self-aware, questioning preconceived ideas and only admitting new knowledge as fact when it had been tested and studied from all angles. Hooke's enthusiasm for instruments was boundless and enduring. At a time when 'mechanic' was a pejorative term with implications about social class (gentlemen were generally unwilling to get their hands dirty with such things), Hooke argued that a good scientist had to be mechanical in order to understand how the mechanisms of the natural world worked.

This may not have appealed to some of his Royal Society colleagues, but his early experience with Boyle's air pump, and his own improvements to the microscope, confirmed to him that instruments would be the key to scientific progress. As we will see, he went on to befriend and work closely with a number of London's most successful scientific-instrument-makers, and he continued showing mechanical devices of his own invention at Royal Society meetings into the 1680s and '90s.

Once the mind and body had been suitably prepared and furnished with mechanical aids, the business of making a 'philosophical history' could begin. Hooke envisaged this history, or collection of facts, as his 'Repository of Materials': an orderly builders' yard stocked with foundation stones and lumber that would, at some point in the future, take its place in the edifice of new knowledge. In order to make a full and useful collection of materials there were six things that it was absolutely necessary for a philosopher to be skilled in, or willing to do. First, he should be well-read in the existing philosophical literature, because by knowing what had been discussed in the past, methods that had succeeded or failed and so on, he would already be used to thinking up queries and hypotheses and discovering errors. Second, he should be skilled in mathematics and mechanics: 'those things, which will most assist the Mind in making, examining, and ratiocinating from Experiments'. Third, he needed to be able 'to design and draw very well', both to help his own thinking and to explain ideas to others. Fourth, he should be able to work patiently and diligently, without being distracted by the 'sweet singing Syrens' of discoveries that might be attractive or lucrative but were not relevant to the main goal of the investigation. Fifth, he should get 'what Help he can from others to assist him in this his Undertaking': without help, a man could spend his whole life on the undertaking and not achieve his design. And finally, he should, 'as Columbus did', share his findings freely and in a form that was easily understood, taking care to acknowledge other people's contributions.[15]

Hooke's emphases here reveal what he saw as his own assets in the philosophical enterprise, and in the following chapters we will see them displayed in his own work. He was a voracious reader and would go on to build up a large library. He was an able mathematician, though not in the same league as his slightly younger contemporary Isaac Newton, and a skilled draughtsman and mechanic. His final three desiderata are also revealing. His allusion to Christopher Columbus was not new: by the mid-1660s, it had become something of a cliché to connect the discovery of the New World by Europeans and the discovery of new fields of knowledge. Both were seen as heroic ventures, enlarging the territory of European domination and giving new wealth and power to European states. Hooke did not consciously invoke the imperial aspect of the metaphor, although of course it is still present. Instead, rather characteristically, he emphasized the more mundane parts of Columbus's endeavour: the fact that he had been supported with 'Ships, and Men, and Money' (experimenters would also need supporters, probably wealthy ones); that he had not allowed himself to be distracted from his course by 'Syrens' or anything else, but had used signs such as shallower water, weeds drifting on the surface of the sea and white clouds on the horizon to guide him on his way; and he had not attempted to keep his discoveries to himself.[16] Here, as elsewhere, Hooke concentrated on the practical aspects of the scientific method – funding, logistics and communication – while at the same time appealed to his audience with the mental image of the scientist as adventurer and explorer of new worlds.

Having described the ideal experimental philosopher's attributes, Hooke moved on to explain how they should go to work. This was the most detailed part of his scheme, and it will be explored further in the chapters that follow. To begin with, Hooke emphasized some important basic points. He explained that scientific study would need to encompass the whole universe, and therefore nothing should be left out. This would involve a reversal of common ways of thinking: for science, the things seen as 'the most trivial and vile' in the eyes of the world were going to be more significant than the most precious and exotic artefacts. In order to ensure that everything was explored, he suggested a way of breaking down the universe into smaller themes: 'the History of Comets and Blazing Stars', 'the History of Smells', 'the History of Birds' and 'the History of Masons, Stone-cutters, Statuaries, Sculptors, Architects &c' all had their place in his complex taxonomy of topics.[17] When the philosopher had identified a theme and had listed a set of sub-questions relevant to the theme, it was time to consider the best way of attacking the work. For example, whether 'Histories and Observations from abroad' would be needed, or 'Experiments, Observations and Tryals at home', and how far the senses could work unassisted, or whether they needed the help of instruments or other 'Artificial Helps'.[18]

It was this combination of the human senses and assistive technology that Hooke believed was the key to advancing knowledge. First, all possible information must be gathered by the senses unassisted – this primary data would catalogue the 'more obvious and superficial Proprieties [properties] of Bodies'. Then the relevant artificial helps should be used. These 'helps' were intended to assist the senses either by refining and processing the initial data or by extending the capabilities of the human faculties, allowing more data to be collected. The former required 'Standards and Measures' for the exact description or measurement of phenomena: for example, standards that described degrees of heat or cold, degrees of dryness and moisture in the air, degrees of light and a standard way of determining colours. The latter involved extending the sight and hearing by using scientific instruments such as the microscope, telescope or otocousticon (hearing trumpet), as well as a range of techniques that made it possible for the senses to perceive qualities that would otherwise be too faint or too strong, extending the senses of smell, touch and taste.[19]

This data-gathering stage had already begun; perhaps Hooke had the Royal Society's meetings in mind when he described the subjects that required investigating and the instruments needed to observe and measure. But he saw his own role differently. These discoveries made by the senses were like the letters of the alphabet in the process of learning to read and write. They had to be known and understood, and some were significant enough to be registered, but the final necessary part of the process was induction. Here the philosopher needed to stand back from the information provided directly by the senses, and 'by putting of several Observations and Informations together, and collecting from them, and by reasoning and deducing from them', move on to new experiments that would drive forward the inquiry.[20]

Hooke admitted that his process of reasoning seemed on the surface to be nothing more than 'what every Man would do'. And yet, he argued, it had a key advantage. The common train of thought trusted to the 'bare' power of the senses, memory and understanding. But by using Hooke's method, all these faculties were reinforced:

> the Senses are helped by Instruments, Experiments, and comparative Collections, the Memory by writing and entering all things, ranged in the best and most Natural Order; so as not only to make them material and sensible, but impossible to be lost, forgot, or omitted, the Ratiocination is helped first, by being left alone and undisturbed to it self, having all the Intention of the Mind bent wholly to its Work, without being any other

> ways at the same time imployed in the Drudgery and Slavery of the Memory.[21]

At its heart, Hooke's scheme was simply a way to ensure a methodical approach to a question and an ordered set of results. Some aspects stand out, however, and would have presented a challenge to his contemporary audience. He rejected the common idea that some natural processes were impossible to detect or understand. If nature was mechanical, he argued, then understanding simpler operations would lead to understanding of the most complex operations. His recurring emphasis on the mechanical was fundamental to his concept of nature, but he had to tread carefully: as we will see in Chapter Seven, his Royal Society colleagues objected to his description of the relationship between the mind and the soul, which they saw as overly mechanistic. They would have been more comfortable with Hooke's general insistence that collaboration was necessary for scientific progress, but many did not share his enthusiasm for discussions with craftsmen and labourers about the specifics of their trades.

By the mid-1660s Hooke had come a long way in his scientific career since starting work for Robert Boyle a decade previously. Perhaps what prompted him to write at greater length about a scientific method was his unusual situation. In a society where most men and women followed their parents' footsteps into a trade, or trained for one of the professions, Hooke's position was out of the ordinary. As the first salaried English research scientist, he was unique. By writing about his scientific method, Hooke could hopefully induct others into his own position: he was, essentially, creating a professional identity for himself. It was an identity that was strongly bound up with the city in which he lived and worked.

TWO

A City Where All the Noises and Business in the World Do Meet

In the summer of 1665, shortly after Hooke had installed himself in Gresham College and begun his work for the Royal Society, the plague began to rage through London. Those who could left the stricken city for the cleaner country air. Hooke went with his Royal Society colleagues John Wilkins and William Petty to Durdans, near Epsom, the country house of George Berkeley FRS, later first earl of Berkeley. They were visited there by John Evelyn, who found the three men busy 'contriving Charriots, new rigges for ships, a Wheele for one to run races in, and other mechanical inventions', adding admiringly that 'perhaps three such persons together were not to be found else where in Europ, for parts and ingenuity'.[1] Like Hooke, Wilkins and Petty were inventive and mechanically minded: Petty had recently designed his 'double-bottom' boat, an early European twin-hulled boat, and Wilkins's 1648 book *Mathematical Magick; or, The Wonders that May be Performed by Mechanical Geometry* dealt with flying chariots and submarines as well as more traditional mechanical devices.

If we view Hooke's philosophical method as an attempt to write his professional identity as a scientific researcher into existence, the section about requiring help from others assumes extra prominence. It starts with the frequently made point that one person acting alone will not be able to add very much to the sum of knowledge, which seems to suggest the need was for scientific collaborators. However, it immediately turns to financial support, describing exactly Hooke's own situation: 'there is much of Expence requisite [for natural philosophy], which every one cannot so well bear, that may perhaps be otherwise fit for this Employment'. Hooke's stipulation was that the budding philosopher 'must therefore here also imitate Columbus, endeavour to be provided with Ships, and Men, and Money', and anything else necessary for the business at hand.[2] The idea that the successful experimental philosopher also had to be a successful fundraiser would not have shocked Hooke's Royal Society colleagues, who had similar intentions but whose attempts to woo Charles II and other wealthy aristocrats would never (in this period) prove to be particularly fruitful. Nevertheless, Hooke's funding from Sir John Cutler showed that it was theoretically possible to be provided with money, if not ships and men. Gresham College was based in the heart of the City of London: the financial hub of the Royal Exchange, the goldsmiths of Lombard Street, the wealthy livery companies and the Guildhall and later the Bank of England were only a few streets away. Hooke may have seen opportunities in these city connections that his courtier colleagues were slower to spot.

Hooke quickly became a confirmed Londoner. His plague-prompted country visit was an unusual departure: from the early 1660s onwards his whole social and scientific life was focused on London. Occasionally he was enticed out into the country to oversee work on stately homes he had designed, but his diary does not give the impression that he enjoyed these excursions very much. He returned from a couple of days at Shenfield in Essex 'with a twittle twattle company in the coach', and on the Friday after spending a whole week away in Berkshire and Wiltshire he recorded a relieved 'at home by 11. Deo Gratias. [Thanks be to God.] Drank sider at Goodwins.'[3] In September

1676 Hooke rode out of London with his friend and Royal Society colleague Sir John Hoskins to Banstead Downs in Surrey. The goal of the outing was to investigate an echo that could be heard in Banstead's deep well, and the trip was only marred for Hooke, an infrequent rider, when he was 'hurt in testic[le] by jolting horse'. On the way home they observed 'the cloud of smoke over London', which Hooke estimated to be a 'half mile [800 m] high and above 20 miles [32 km] long'.[4] Living inside the city's grubby environs, he was not normally able to observe the smog rising from its industry and cooking fires. Even when his brother John committed suicide on the Isle of Wight in February 1678, rather than going in person Hooke sent his assistant Tom Crawley to the island on a borrowed horse with 40 shillings. He himself stayed in London, gaining audiences with the king and the secretary of state to try to prevent John's estate from being seized by the crown, as would normally have happened after a suicide.

Hooke had not started his diary in September 1666, so we do not have his own first-hand account of the Great Fire. For his more poetic fellow diarist John Evelyn FRS, the flames conjured an image of apocalyptic devastation, 'Sodome, or the last day', the smoking ruins 'resembling the picture of Troy' in the aftermath of the Trojan War. Miraculously, Gresham College was not burnt, but the experience must have been terrifying for Hooke and the other inhabitants. Fire was a relatively common occurrence in London at the time, though mercifully not on the same scale as the Great Fire, and Hooke did record other outbreaks in his diary. Ten years later, in May 1676, he and his niece Grace observed 'the great fire in Southrick which burnd 8 or 900 houses'. Ironically, the vantage point from which they watched the flames across the river in Southwark was the Monument, a pillar commemorating London's fire on Fish Street Hill, which Hooke was still completing.[5]

We may not have Hooke's verbal response to the fire, but we do know that he quickly proposed a plan for rebuilding the city, apparently in a grid format.[6] His Royal Society colleagues Christopher Wren and John Evelyn also presented their own plans separately to the king. This was clearly an opportunity for the new philosophers to demonstrate that their work had the potential to be useful for their fellow citizens. Hooke's proposal was preferred by the city authorities to whom it was submitted, but in the event, none of these visions for a statelier, better-proportioned London came to fruition. The most urgent need was to start rebuilding. Perhaps because they had been impressed by his plan, in March 1667 city officials appointed the otherwise totally inexperienced Hooke to act as one of the surveyors tasked with measuring and marking out plots of land so that building could commence. This was a laborious and complicated task, and one that had the potential to involve Hooke and the other surveyors in endless disputes over boundaries, obstructions to access, drainage and light, building regulations and adequate recompense for land taken by the authorities in order to widen roads or make other improvements. Nevertheless, Hooke took on the work alongside his positions as Royal Society curator and Gresham College lecturer.

Hooke's initial work as surveyor involved staking out roads, and then staking out and measuring the foundations of building plots and certifying his measurements in writing for those who were ready to rebuild. Some streets were ordered to be widened so that carts bringing commodities up from the River Thames could pass along them more easily. This of course meant that landowners on either side of the street lost some ground, which inevitably led to complaints. No one was allowed to begin rebuilding until one of the city surveyors had viewed and certified the ground, and new building regulations controlling, among other things, materials, party walls, balconies and gutters were put in place to try to ensure that the city would be a healthier place to live in future and would never again suffer such a catastrophic fire.[7]

The rebuilding work was still going on in the 1670s, and Hooke's diary attests to his continuing activities on behalf of the city. Christopher Wren had been appointed as surveyor-general of the king's works in 1669, and the two men worked closely together on many of the city buildings. Hooke directed work on a number of the city churches, which were slowly being rebuilt under Wren's supervision, but he also undertook various commissions for city guilds. He was still overseeing the finishing touches of the 'Fishstreet piller', or Monument, noting with frustration on 10 April 1679 a mistake made in engraving one of the inscriptions on the column: 'Knight cut wrong. R for P.' Hooke worked with a similar team of bricklayers, carpenters, stonemasons and plasterers on many of his commissions. Thomas Knight had been appointed as city mason in 1667, and he worked with Hooke periodically, most notably in 1677 when Knight was employed to do some work on the conduit at Snow Hill, Holborn, part of the system that provided drinking water to Londoners. The two men surveyed the site together in August, although Hooke's confidence in Knight may have been impaired by having witnessed him 'Drunken' at a local alehouse just two days after Hooke had delivered to him the design of the conduit.[8]

The Snow Hill conduit was close by the site of one of Hooke's other major projects in this period. In 1671 work commenced to make the River Fleet navigable at high tide up to Holborn Bridge, and to build quays and warehouses by the side of the New Canal, as it was known. The river itself was almost exclusively referred to by Hooke and his contemporaries as 'Fleet Ditch', which gives a good sense of its condition at the time. Hooke brought all his usual energy and focus to the task of removing the silt and rubbish that filled the Fleet, widening the canal and shoring up the banks, but it must have been a somewhat morale-sapping project. The records of the City Lands Committee, the body responsible for the work, show that the project turned out to be much more difficult (and expensive) than had been envisaged, and Hooke and his workmen suffered numerous setbacks. In a particularly grave incident in January 1674, a large section of wall dividing the Fleet from the garden of Alderman John Nicoll collapsed suddenly into the channel, killing several labourers and injuring others. Hooke and his colleagues were ordered to assess the damage and report on the probable cause, which they concluded to be a combination of poor foundations partly washed away by 'the Extream wetness of the season' and too much strain on the wall.[9] Hooke was more direct in private, recording in his diary that Nicoll's wall had been 'thrown downe by his own potashes etc'.[10] (Nicoll was a soap-boiler, a trade that required potash made from burnt plant material.)

Despite its difficulties, Hooke's work on the Fleet channel provided opportunities to assess the benefits of different materials and techniques, and sometimes to try out new inventions. The apothecary and inventor John Conyers demonstrated a new pump that he claimed was twice as effective as standard pumps of a similar size (illus. 5). His 'engine' gained the approval of Dr Francis Glisson and Dr Jonathan Goddard, both fellows of the Royal Society, and a full description was printed in the society's journal. Hooke later had protracted discussions with Sir Robert Southwell, landowner and president of the Royal Society, on the subject of sluices, quays and wharves, passing on knowledge gained during his work on the Fleet. Even in the midst of his fraught investigation into Alderman Nicoll's wall, Hooke had time to observe interesting things, recording an aside in his diary: 'I found in alderman Nichols garden a Mulberry tree that had formerly been splitt rivetted together by an iron

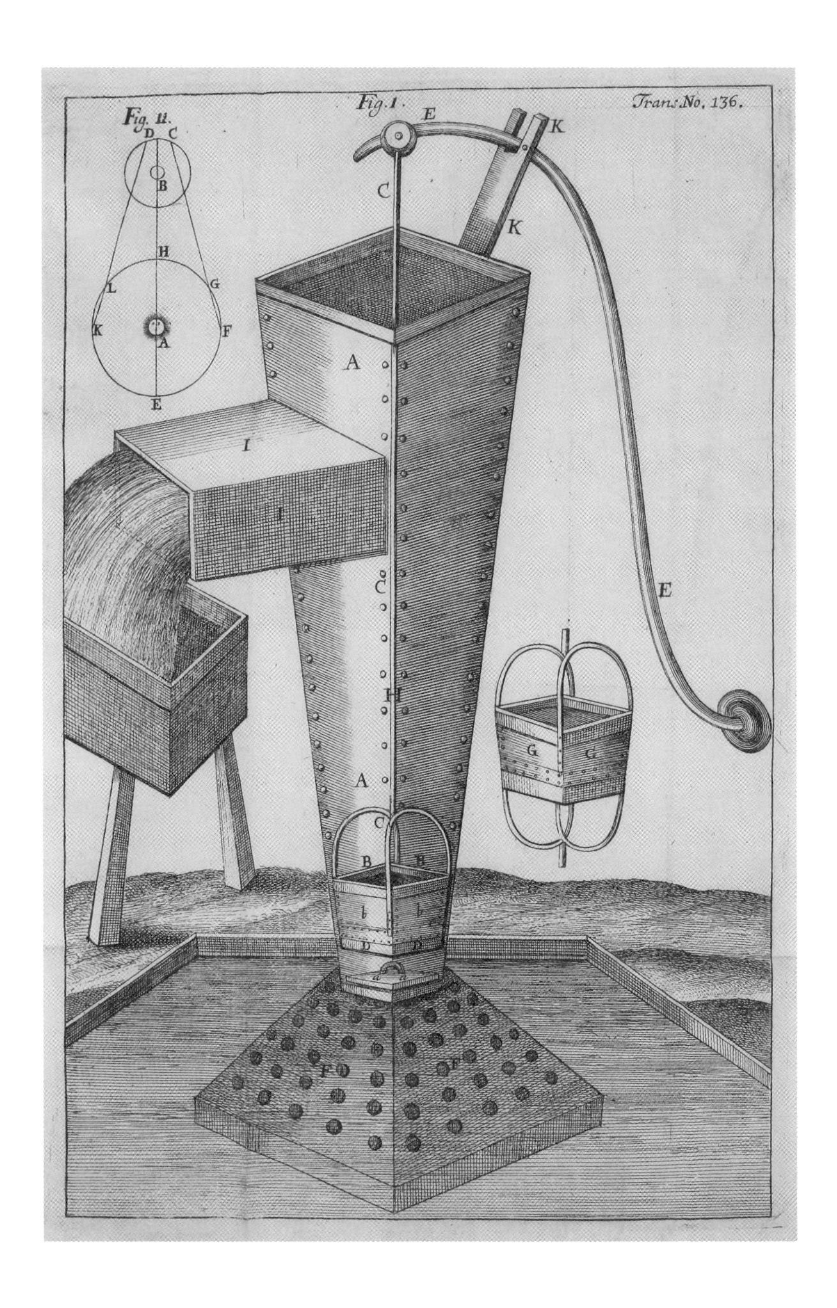

5 Improved water pump designed by John Conyers, illustrated in *Philosophical Transactions of the Royal Society* (1677).

bolt ¾ inch [20 mm] big. the tree thrived much and ever since bore vast quantitys of mulberys and never before any. Let more tryalls hereof be made.' There is no further evidence of experiments with mulberry trees, but other opportunities proved to be more influential on Hooke's thinking. During the development of St James's Square in the early 1670s, brickmakers had sunk several wells on-site, around 6 metres (20 ft) deep, to provide water for making bricks. Hooke recalled later that he had gone down into several of them and found at the bottom 'a Layer of perfect Sea Sand', studded with seashells, bones and other substances, a sample of which he had collected to show the Royal Society. He used this evidence of shells found far from the sea, in places where they could not possibly have been deposited 'by the Industry of Man', to support his theory that the earth had undergone constant changes during its history.[11]

By the mid-1670s Hooke's architectural commissions were expanding, dominated by his work on a grand mansion for the courtier Ralph Montagu in Bloomsbury, on the site of what is now the British Museum. At the time it was a pleasantly rural location, on the edge of the town, backing on to fields. Hooke designed the house and oversaw the building closely. One of the largest residences in London, Montagu House was designed to display Montagu's wealth and taste. John Evelyn visited in October 1683 and described it as 'a fine palace, built after the French pavilion way'; he was a noted art connoisseur, and thought that the Italian painter Antonio Verrio's frescoes decorating walls and ceilings were 'comparable certainely to the greatest of the old Masters, or what they so celebrate at Rome'.[12] Unfortunately Hooke's original house burnt down in 1686 and had to be rebuilt. In this period Hooke also designed and directed the building of Bethlehem (known as Bethlem, or Bedlam) Hospital, at a site in Moorfields just outside the city wall. Hooke designed an enormous and extremely grand building for the governors

overseeing the project, with 136 cells for inmates opening off galleries running the length of the (very long) building (illus. 6).[13] Another major institutional commission saw him designing and building premises for the Royal College of Physicians in Warwick Lane, London. This set of buildings included a theatre in which anatomical dissections could be performed and lectures delivered (illus. 7). His work on these building sites, private and public, must have made Hooke a familiar figure in Restoration London. He was out almost every day, walking through the city and newly developing streets and squares to the west in Soho and Bloomsbury, across to Westminster for conversations with Christopher Wren at his house in Scotland Yard and then back to Gresham College, sometimes travelling by coach or by water on the Thames, but most often on foot.

Despite Hooke's own emphasis on his scientific work as the guiding principle of his life, his diaries tell a different story, with many more references to his work as city surveyor and architect than his duties for the Royal Society and as Cutlerian lecturer or Gresham College professor. This may simply be because he recorded the details of his scientific life elsewhere. Nevertheless, we cannot ignore the fact that his surveying and architectural

6 Engraving of Bethlem Hospital by Robert White, 1677.

work brought him prominence, and a degree of wealth beyond that which could have been expected from his scientific activities. He became a respected figure in the city's circles of aldermen, merchants, guild officials and tradesmen. In 1672 he was elected as a governor of Christ's Hospital school, a charitable institution for the education of poor children. Because the school was a city institution, governors were elected by the Court of Aldermen, the main administrative body of the City of London. Many wealthy and influential citizens were governors, as well as some of Hooke's Royal Society colleagues. In 1673 Charles II established the Royal Mathematical School as part

7 Engraving of Hooke's anatomy theatre at the Royal College of Physicians by David Loggan, 1677. Inside, steeply raked seating and plenty of light from the windows and cupola made it easier for attendees to see the anatomical dissections.

of Christ's Hospital, intended to train boys in mathematics and navigation to prepare them for entering the navy. There was some suggestion that the boys from the Mathematical School would attend Hooke's Gresham College lectures in geometry, and they seem to have done so: coining a new collective noun, Hooke noted an 'alarum of Blew coat Boyes to hear Lecture' in October 1679.[14] (The boys were identifiable by their distinctive blue uniform.) Despite possible mixed feelings about having them at his lectures, Hooke did take an active interest in the school. He designed an allegorical scene and motto for a school badge in 1674, and he contributed to discussions about the appointment of new mathematics teachers.

Much of Hooke's interaction with the city authorities took place at the Guildhall, the administrative hub of the city. Here Hooke met a succession of lord mayors and aldermen, elected from among the city's wealthiest citizens. He was primarily involved with the work of the Committee for City Lands, which was overseeing the process of rebuilding after the fire. Generally, things went smoothly, although occasionally Hooke grumbled in his diary. After a day of frustrations, he wrote, 'Thomson proud fool. Controller kind . . . nothing done in answer of Petition. foold and foppd. Oliver a villaine.'[15] John Oliver was the other principal city surveyor, and therefore was someone with whom Hooke worked closely, although, as this entry suggests, he was not always complimentary towards his colleague.

Nevertheless, Hooke met with the lord mayor and other office-holders regularly, and he counted some of these men among his close and trusted associates. The East India Company merchant Sir John Lawrence, who had been lord mayor in 1665, was elected to the Royal Society in 1673 and went on to serve on the society's council, suggesting that he was genuinely interested in their affairs. He and Hooke met frequently in the 1670s on city business and socially, and they clearly had overlapping interests.

It was at Lawrence's house that Hooke met the 'projector', Nicholas Daintree, and listened to Daintree's proposals for fighting fires in London by means of pipes fed by watermills in the three southern arches of London Bridge. Hooke felt that these plans were 'extravagant', but there was a clear need for more powerful firefighting equipment. Conversations between Lawrence, Hooke and Daintree continued, and despite a further dismissive comment from Hooke – 'nothing worth a farthing' – he seems to have collaborated on Daintree's 'engines'. The Guildhall administration later agreed to pay Daintree £20, a large sum of money at the time (more than a year's wages for an average earner), to reimburse him for money spent on 'Engines and pipes and other things'.[16] Hooke and Lawrence also requested payment from the city to John Tinker, whom they had directed 'to make two moddels for an Experiment that may be tryed with a pound of Powder for Each'.[17] Experiments with gunpowder were commonplace in the period: Hooke had performed some at a Royal Society meeting a few months earlier. By linking this scattered evidence from the city's archives with brief mentions in Hooke's diary, however, we can see that Hooke was collaborating on an experimental research programme that ran parallel to his work at the Royal Society but was shaped by different priorities. It also demonstrates the extent to which Hooke was valued by the city administration as a scientific colleague. For example, in March 1675 Hooke was appointed as one of a small group to report back to the Court of Aldermen on the 'usefulnes and practicablenesse' of a new fire engine designed by Mr Lattenhoven, a Dutch man.[18] It was clearly not part of Hooke's duties as city surveyor to perform this kind of task, and yet he was the obvious person to join the nominated aldermen. Brief mentions in his diary suggest that he met Lattenhoven, or the 'Dutch Engineer', as Hooke called him, several times, and it seems that the fire engine was even demonstrated in the courtyard at Gresham College.

These activities illustrate perfectly Hooke's approach to the two sides of his working life, which we might loosely think of as theoretical (Royal Society) and practical (surveying and architecture). This is an artificial division, however: Hooke's writings and actions suggest that he did not see these as two separate spheres, but instead as an integrated whole. If this was the case, what continuity might there have been between Hooke's architectural or surveying work and his scientific method? Perhaps the most immediately striking is his vision of the scientist as an architect. When he discussed the best way to organize and record new knowledge, Hooke almost always used metaphors of architecture. As we have seen, he referred to the data gathered by experiment and observation as the building materials that would be used to construct a new structure of knowledge. Sometimes this was oddly specific. He advised, for example, that the experimental data recorded in a philosophical history should be 'sound and good, and cleans'd and freed from all those things which are superfluous and insignificant to the great Design'; but not so brief as to cut out the 'many little Circumstances' that might be significant when the data came to be used. He explained this further by comparing the recording process with a foreman laying up timber: 'the keeping on a branching part does make it serviceable for many Designs which it would be wholly unfit for, if it had been squared off.' For builders in the period, a naturally curved piece of timber was invaluable for making strong arches.[19]

Hooke's key point here is that like the architect, the scientist needed to think simultaneously about the goal (the scientific problem that needed to be solved) and the collection of materials or data required in order to solve the problem. By comparing an experimental philosopher with an architect, Hooke highlighted what he seems to have seen as one of the problems with the Royal Society's programme. Architecture required planning and intelligent choice: buildings do not simply raise themselves from their materials, they need the controlling mind of the architect at work. Similarly, the experimental philosopher needed a plan: in Hooke's view, strong scientific theories would not simply emerge from a tangled mass of data. Yet his early Royal Society colleagues were in general wary about making hypotheses. In fact, Hooke had to apologize to the society at the beginning of *Micrographia* for breaking their cardinal rule of 'avoiding Dogmatizing [stating ideas dogmatically], and the espousal of any Hypothesis not sufficiently grounded and confirm'd by Experiments.'[20] (Despite agreeing that this was an excellent rule in general, he went on to use the word 'hypothesis' a further 54 times in *Micrographia*, which suggests that he allowed himself a bit more latitude than his colleagues might have approved of.) Later in his career, in the preamble to one of his lectures on earthquakes, he confronted head-on the Royal Society's reluctance to draw up theories, arguing that a more directed approach might lead to better results than the haphazard collecting of random facts. 'When this mighty Collection [of data] is made, what will be the use of so great a Pile? Where will be found the Architect that shall contrive and raise the Superstructure that is to be made of them, that shall fit every one for its proper use?'[21] It may be that Hooke's experience working as a surveyor and architect gave him a different perspective on the scientific endeavour. The architect metaphor describes an active and purposeful method, one that was consistent with Hooke's optimism about scientific progress and his certainty that the right method would produce results. In his building work he saw several complex and lengthy projects through to completion: why could scientific progress not work in the same way?

While they may not all have agreed with him about the benefit of hypotheses, Hooke was not alone among his Royal Society colleagues in considering London a fitting home for science.

This sentiment was expressed most bombastically by Thomas Sprat in his *History of the Royal Society*: 'if we should search through all the World, for a perpetual habitation, wherein the Universal Philosophy might settle it self; there can none be found, which is comparable to London.' London, Sprat argued, had several key advantages over other cities: 'It is the head of a mighty Empire, the greatest that ever commanded the Ocean: It is compos'd of Gentlemen, as well as Traders: It has a large intercourse with all the Earth: It is . . . a City, where all the noises and business in the World do meet.'[22] We should bear in mind that Sprat's 'history' was in fact a piece of propaganda written to persuade contemporaries, and perhaps most importantly the society's royal patron Charles II, that science would be useful. Praising the capital and its inhabitants was part of his attempt to win support, perhaps the kind of financial support that Hooke's method described. Sprat's main point here is that London was a centre of information and mercantile exchange: eventually all news, and all the exotic specimens and commodities that the world could provide, would be channelled into London. For Hooke, though, London offered something literally more concrete as a setting for science. For him, the whole city was a kind of laboratory. We have seen him investigating natural phenomena, descending into wells for samples of fossils, or observing smog from a distance. On another occasion he dashed into a house that had just been hit by lightning in order to find out what damage had been done. The actual fabric of the city provided a different sort of opportunity. When discussing ways to make accurate observations of astronomical objects, he suggested erecting a very long telescope, asserting confidently:

> Nor will it be difficult in this City to find a convenient Building or Tower for the resting the end of the Telescope of a hundred Foot long [30 m] if it be made

> use of; or of finding a good prospect of a far distant Meridional Object in the Horizon . . . And if a fifty or sixty Foot [15/18 m] Telescope be made use of . . . there are Houses enough to be found of sufficient height.[23]

Towers and houses might be used as mounts for telescopes, and other buildings could provide fixed points against which the movement of the stars could be measured. The size of a comet could be described in relation to the tower behind which it passed, a comparison Hooke found so useful that he included an engraving of it in his printed description of the comet (illus. 8). As a scientific instrument, however, the city did not always answer expectations. Hooke and Wren's idea to use the Monument as the tube of a large zenith telescope (that is, a telescope designed to point straight up) turned out to be impractical because carts bumping over nearby streets caused vibrations.

Hooke's belief that, like Columbus, the experimental philosopher needed to find financial backers did not really translate into new streams of money flowing into the Royal Society's coffers – or his own. Although his collaborations with aldermen and other city merchants and tradesmen were fruitful in other ways, even his association with Sir John Cutler soured over time. Despite having begun by sending Hooke 'a half year's salary' as an advance in October 1664, in midsummer 1670 Cutler stopped paying for the lectures that Hooke was still delivering in his name.[24] Hooke's diary in the 1670s records his many attempts to chase Cutler for the arrears, a situation that Hooke clearly found upsetting and wearing. Nevertheless, Hooke's experimental method is striking for the emphasis that he put on looking outward. For him, external experiences – conversations, time spent watching tradesmen at work, the act of touching, smelling and tasting different substances – underpinned natural philosophy. It was only through close attention to both the natural

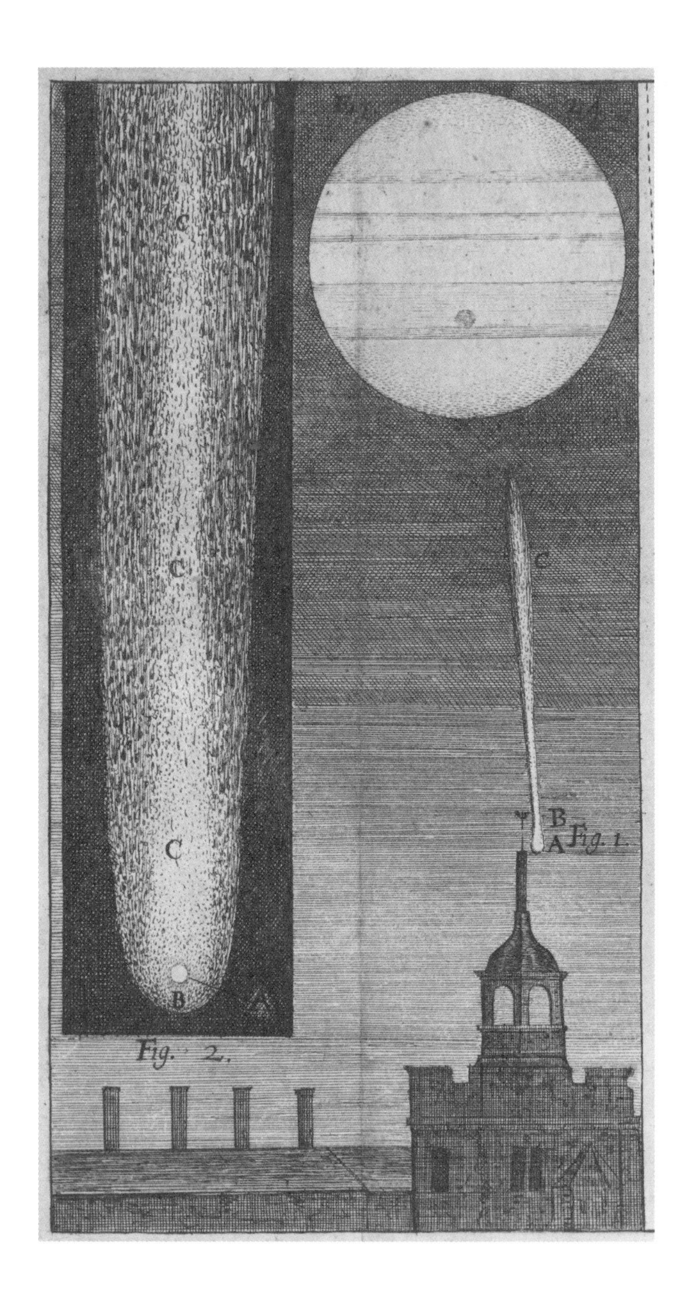

8 Hooke's illustration of a comet observed in 1677, and the planet Jupiter (top right).

world itself and the ways in which it was manipulated by humans that a philosopher could come to understand how things work. His method relied on people of all 'estates and conditions', great and humble, sharing their own knowledge and experience with a philosopher who was willing to listen to everything they had to say and to take their words seriously, even on the commonest of subjects. Indeed, Hooke suggested that when considering a new topic about which he knew little, a philosopher would do well to think about the questions he would ask if he were to meet someone more knowledgeable about the subject. These questions were only intended to prompt ideas about what experiments or observations might be necessary to answer them, but the fact that Hooke used a conversation between a philosopher and an expert practitioner as a way to frame his preferred method of investigation is telling. One thing that London provided in abundance was opportunities for conversations.

THREE

Much Love and Service to All My Friends

On 31 December 1676, at the end of a busy year, Hooke sat down to tally his accounts. He calculated the amounts that he was owed in unpaid salary from his various employers: £325 from Sir John Cutler (for six and a half years' worth of Cutlerian lectures at Gresham College); £450 from the City of London for three years' salary as surveyor; £150 from Sir Christopher Wren for work on the city churches; and smaller but still significant amounts from other employers. In total he was owed more than £1,300, a stupendous amount at the time. After his death, almost thirty years later, contemporaries marvelled at the large amount of cash found in Hooke's rooms. In the final years of his life he had spent very little on his day-to-day necessities, leading some to write mockingly about his miserly ways. It seems likely, however, that this experience of not being quite sure when (or if) his salary would be paid may have weighed on him in later life.

Compared with this vast sum, the amounts Hooke owed to various tradesmen on that December night in 1676 were tiny: 12 shillings for a load of coal, 12 shillings for wine and an unspecified amount for two pairs of shoes and goloshes. The largest sum was £5 for a velvet coat and lining. Sandwiched between debts to his wine merchant and to a couple of instrument-makers for his new quadrant, Hooke has written 'much Love and Service to all my friends I owe'.[1] It is a strange and sudden shift of focus in the middle of his list of accounts, but the diary as a whole bears out Hooke's acknowledgement of this debt to his friends. Hooke did conduct high-profile arguments with several of the leading philosophers of the day, including Isaac Newton, Christiaan Huygens and Johannes Hevelius. This led some early commentators to characterize him as bad-tempered, but more recent writers have demonstrated the legitimate concerns that led to these public disputes.[2] His diary, however, depicts a gregarious life. Setting aside purely work-related meetings, almost every entry bears evidence of tea- or coffee-drinking, chess playing, browsing in bookshops, walking or conversation with colleagues, chance-met acquaintances and friends.

Clearly Hooke enjoyed being with other people: no one naturally introverted could sustain such a varied and constant mix of social interactions. As we have seen, however, he also regarded collaboration as one of the fundamental requirements for experimental philosophy. This could take different forms, such as having a co-worker present when doing experiments. He suggested that a philosopher making an experiment or observation should

> get some such Person to be present as has not been acquainted with Experiments on that Subject, though ingenious and inquisitive in other Physical Searches, because such a Person may take notice of many Particulars which are in themselves very observable, but were and would still have been neglected because of their being obvious, and because the Inquirer having been long accustomed to the seeing of them without thinking them any ways considerable, will be now very prone to do as formerly, slight and neglect them, and this will be instructive to him in shewing him in what things he is most likely to be overseen in, as well as shewing him the things themselves.[3]

It is clear from this description that Hooke was not simply thinking of a laboratory assistant helping with the experimental apparatus or procedures. He recommended having a person who was an expert in a different area; that is, an equal who would bring a different point of view to the material being examined. Perhaps Hooke's own experimental collaborations with Robert Boyle had demonstrated how useful it could be to work as a team. The idea that an experimenter might disregard some phenomena because they were too obvious accords with Hooke's recurring point that common and ordinary things were apt to be overlooked but were just as likely to provide useful information as the strange and exotic. Interestingly, Hooke also points out here that the process of making and discussing an experiment could indicate something about the experimenter himself (by highlighting the things he was most likely to miss), as well as the results of the experiment. It fits with Hooke's insistence in his philosophical method that the mind as well as the body of the observer needed to be prepared for making observations.

If Hooke did see echoes of his relationship with Boyle in this co-working situation, he had an opportunity to replicate it in his own experimental life by taking on a series of young men who acted as his live-in assistants, or apprentices. Henry Hunt was one such apprentice. He came to Hooke after the death of John Wilkins, bishop of Chester, in whose employment he had initially been. Wilkins was a mentor to Hooke, Christopher Wren and others, and his death in November 1672 was much lamented. Hooke noted in his diary on the day of Wilkins's death, 19 November, 'a conjunction of Saturn and Mars. Fatall Day.'[4] Saturn and Mars were both believed to have malign influences on human affairs, so their alignment was particularly ominous. This uncharacteristic attention to the astrological betrays Hooke's emotional response to Wilkins's death. The philosophical community rallied around to provide for Wilkins's household. A few days later, Hooke met with Lord Brouncker, the Royal Society president. Hooke told Brouncker the news of Wilkins's £400 legacy to the society, and was asked to take on Wilkins's man, Henry Hunt; Christopher Wren took on another of Wilkins's servants, Benjamin Janeway.

Hunt was perhaps an even more valuable asset to the society, and certainly to Hooke, than Wilkins's £400. He quickly became Hooke's right-hand man, running errands, looking after the Repository, assisting with experiments, sketching natural history artefacts and, in time, engraving plates for Hooke's publications. In 1676 Hooke suggested to Hunt that he might want to take the place of one of Wren's architectural assistants who had recently died, but he refused, instead taking up an official post alongside Hooke at the Royal Society as 'operator'. His loyalty to Hooke (and indeed the Royal Society itself) remained constant. Later in his life, 'HH tea' was one of the most frequent notes in Hooke's diary, and it has been pointed out that Hooke always treated Hunt more like a son than a paid assistant. Hooke's other assistants never quite occupied the same position. The most promising seems to have been young Tom Gyles, the son of Hooke's cousin Robert. Tom had been sent up to London from the Isle of Wight and according to Hooke was 'good at Reading Arithmetick etc' and wanted to go to sea.[5] Hooke began teaching him some of the basics of navigation, and despite occasional grumbles about Tom's laziness, his diary entries reveal his distress when Tom died of smallpox in September 1677.

Other kinds of collaborations were no less important to Hooke. Conversations with people from all walks of life provided a constant stimulus for new ideas and an opportunity for him to talk through his own train of thought. Hooke and his friends met regularly in coffee-houses, which had become ubiquitous in Restoration London (illus. 9). In the early 1670s Hooke favoured Garraway's coffee-house, run by Thomas Garraway at

premises in Exchange Alley near the Royal Exchange. In 1677 he began to patronize Jonathan's, also in Exchange Alley. The Royal Exchange, newly rebuilt after the fire, housed merchants and shops and was a popular place to go for the latest news. The nearby coffee-houses were equally popular. At Garraway's and later at Jonathan's, Hooke drank chocolate and coffee, and often smoked a pipe of tobacco. Sometimes coffee disagreed with him, making him cough or giving him a 'stoppage of stomack'. He gave up smoking tobacco several times but always took it up again, seemingly for its medicinal properties ('Cleerd all with Retaking Tobacco', he wrote, and 'mightily refresht by tobacco').[6] Hooke's mid-1670s shift in patronage seems not to have upset Garraway, whose coffee-house was perhaps too popular for the loss of Hooke's custom to worry him. He also seems to have regarded Hooke as more than simply another customer. In 1677 he consulted Hooke about a funeral monument for his daughter, perhaps Susanne, who had been baptized in 1667. Hooke had

9 Anonymous, *Interior of a London Coffee-House*, *c*. 1690–1700, drawing on paper. Papers on the tables might contain the latest news and satirical attacks on the government or members of the royal court.

attended her funeral in November 1676, seemingly as an honoured friend of the family: he noted in his diary that he had been given a mourning ring, a 'favour' and a pair of gloves. Rings and gloves were frequently given to relatives or close friends of the deceased at funerals as mementoes. Hooke discussed his design with Garraway and then commissioned a woodcarver to produce the monument in a local church.

Coffee-houses provided a neutral space where like-minded men could meet to talk, exchange news and gossip, read the papers, conduct business, drink tea, coffee or chocolate and smoke. During the 1670s and '80s Hooke frequently met an inner circle of close friends at Garraway's or Jonathan's. Sometimes this was by prior arrangement, for example meetings of the several philosophical clubs he and others instituted during this period. One of these was the club for 'natural philosophy and mechanicks' devised by Wren and Hooke and begun on 1 January 1676. All the members were fellows of the Royal Society, but prompted by his dissatisfaction with the society at the time, Hooke noted that the club members 'Resolved upon Ingaging our Selves not to Speak of any thing that was there reveald sub sigillo to any one nor to declare that we had such a meeting at all.' It says much for Hooke's commitment to collaborative research that his response to problems with the Royal Society was not to withdraw and work in isolation, but to form a new group of people to work with, albeit one where he felt he could control the flow of information outside the group more effectively. Hooke himself announced 'sub sigillo', under a seal of secrecy, that 'all plants were femalls', that they were fertilized by insects and that 'there was upon every great plant a herball of smaller plants microscopicall growing on them and a heard of various animalls microscopicall ranging among them'.[7]

Like Hooke's herd of microscopic animals, the conversation ranged over some fertile ground. Hooke and Wren began by discussing the properties of light, sparked by Isaac Newton's recent letter to the Royal Society on the topic, which Hooke felt had plagiarized his own work in *Micrographia*. Hooke shared details of his new selenoscope, a telescope designed for observing the moon that he said had given him a 'strangely cleer' view of the lunar surface. He argued that there was water on the moon, and that observers on earth could see the bottom of lunar seas as well as the tops of mountains, using as evidence his own experience of seeing the seafloor from the top of a cliff. Wren disagreed, maintaining that there was no water on the moon. This led on to discussions of watery places, how to cross bogs using floats and different methods of walking over ice. From the properties of ice, the conversation moved to fire, including a recollection of 'the fellow that held the Red hot iron in his teeth seen by the Royal Society'. Hooke and Wren discussed ways of making lightning and thunder effects in a theatre, and Hooke explained his theory of lightning (that it consisted of 'subterraneous steams' raised by the air's warmth and 'kindled above'). Wren and Abraham Hill discussed the effects of gunpowder and how far different ordnance could be shot. The conversation turned to seeing in the dark and different forms of bioluminescence. Following Hooke's comment about plants being female, others objected that plants were male and female. Edmund Wylde described his experiments on plants grown in sealed containers. This brought the conversation back to insects, and Wren recalled John Wilkins's experiments demonstrating that no insects would appear on meat that was kept covered. Finally, the company dispersed at nine in the evening, with Hooke mentioning on parting his 'new way for the orthography of musick to be discoursed more next meeting'.[8]

This was the longest diary entry that Hooke ever wrote, and the following day he began a 'new Journall of Club'; if he did keep further notes of club meetings, which is very likely, we no longer have them. This and other philosophical clubs brought Hooke's friends together in a way that closely resembled the Royal Society meetings. Each member contributed information culled from his own reading, experiments and observations, or simply gained from day-to-day life. At this meeting, Christopher Wren seems to have made more comments than other attendees; or perhaps Hooke simply recorded more of Wren's contributions because he believed them to be worth noting. Wren and Hooke often discussed their scientific work with each other, even when they were not directly collaborating on problems associated with architecture or civil engineering. Hooke's inner circle of coffee-house companions in this period included Wren and a number of other Royal Society fellows. Abraham Hill was the society's long-standing council member and treasurer, and briefly its secretary. The son of a prosperous merchant, he rented rooms at Gresham College and took an active interest in natural philosophy without being an experimenter himself. Edmund Wylde MP collected natural and man-made curiosities, books, maps, prints and paintings (a contemporary described his house as 'a sort of knick-knack-atory'). Plants and their growing conditions were a particular interest of Wylde's, who was said to have 'sowed salads in the morning to be cut for dinner' and, according to Hooke, claimed that he could prepare soil in such a way that it would produce pea plants without any peas having been sown into it.[9] The lawyer and MP Sir John Hoskins discussed reforming the Royal Society with Hooke, and advised him on his ongoing payment dispute with Sir John Cutler. Hoskins shared Hooke's love of books, and together with Hill they attended in June 1675 an 'atheisticall wicked play' by Thomas Shadwell (probably *The Libertine*). Other close friends included the courtier Thomas Henshaw, Wren's brother-in-law William Holder, mathematicians John Pell and Sir Jonas Moore, physician Daniel Whistler, antiquarian and biographer

John Aubrey, translator Theodore Haak and the merchant and linguist Francis Lodwick. Lodwick was somewhat unusual in that he occupied a different social class from most of the other Royal Society fellows. He traded on the continent, importing cloth and other goods into London from various European cities. His interest in language, however, at some point brought Lodwick into contact with Hooke's mentor John Wilkins, and probably through him, Hooke. At first glance, he seems a strange choice to become one of Hooke's closest friends: sixteen years older than Hooke and lacking a university education, he seems not to have pursued any natural philosophical research. Yet by the 1690s this unlikely friendship, begun in the early 1670s, had developed to the point where Hooke and Lodwick met daily.

Hooke's diary in the late 1680s and early 1690s shows that he was less busy with surveying work, and it was not uncommon for him to stop in at Jonathan's coffee-house twice in one day. His circle of intimates had changed somewhat by this time. He still spent a great deal of time with friends who were also Royal Society fellows, in particular Lodwick, Richard Waller and his brother-in-law Alexander Pitfeild, and he still met Sir John Hoskins regularly. Waller and Pitfeild were more than thirty years younger than Hooke, but they shared his intellectual interests, and Waller would later act as Hooke's posthumous editor and first biographer. The group that met regularly at Jonathan's was more diverse. It included the merchant John Ashby, who worked for the Royal African Company and was thus probably involved in transporting enslaved people from Africa to the Americas. In 1681 Ashby was granted approximately 810 hectares (2,000 ac) of land at Cooper River in South Carolina, where his son seems to have settled. Ashby provided Hooke with information about America, passing on letters from correspondents there who described exotic plants such as the nutmeg tree and 'vines 8 inches [20 cm] diameter 500 foot [152 m] long growing in the swamps'. In 1690 Hooke visited Ashby's house to see a mummy in a case decorated with hieroglyphics and brought home 'a billet of Carolina Laurell' (perhaps *Prunus caroliniana*). The group also included Samuel Meverell, a merchant trading to Russia, Mr Martin, seemingly a jeweller, and John Godfrey, the clerk of the Mercers' Company, among others. Hooke's diary entries are often silent about the topic of conversation when these friends met, presumably meaning that nothing he considered important was mentioned. On other occasions, however, the group discussed topics that would not have been out of place at one of Hooke's club meetings: for example, one evening in April 1693 the conversation ranged over architecture, the Etruscan king Lars Porsena (the design of whose legendary tomb Hooke and Wren had discussed almost twenty years earlier), an 'Arab Library', ancient and modern learning and the teachings of Pythagoras and Aristotle.[10]

Alongside reading, playing chess with his friend Theodore Haak FRS and occasionally attending a play or walking with friends, these meetings were Hooke's main leisure activity. His diary gives the impression, however, that he never stopped being a natural philosopher, even during convivial evenings with friends. There are notes recording conversations on every conceivable topic at Garraway's, from a discussion of spirits and spectres with Wren, to German metal-working practices, to the rings in oak trees, to dubious medical advice (one of the more benign examples is 'Rosemary in the bottoms of ones feet is a present remedy against the Cramp as also snakes skins'). Other chance-met acquaintances benefited from Hooke's own skills: one evening in January 1676 he met 'a tall soldier of fortune' at Garraway's and taught a Quaker 'to make cantilevers'. However, there was perhaps a drawback to having his haunts well known by fellow Londoners. In December 1688 Hooke noted that he had encountered a 'Perpetual motion man' at Jonathan's who 'would not be answerd'. Schemes promising perpetual motion were presented to the Royal Society fairly regularly, and it seemed that this was a particularly confident proposer.[11]

Not everyone approved of Hooke's coffee-house conversations. John Flamsteed, first astronomer royal, complained privately that Hooke 'makes questions to those hee knows are skilfull in them, and theire answers serve him for assertions on the next occasion'.[12] That is, Flamsteed regarded Hooke as someone who pumped his knowledgeable acquaintances for information and then pretended that the ideas were his own the next time the subject came up. It is true that Hooke's philosophical method advised preparing a list of relevant questions on a particular topic by thinking about what one might want to ask an expert in that field: perhaps this advice was drawn from his own experience. Hooke was certainly known for questioning inventors closely about their new contrivances, a habit that irritated and alarmed some of his philosophical colleagues. When the German philosopher Gottfried Wilhelm Leibniz visited London in 1673 he showed the Royal Society his new 'arithmetical machine', a gear-driven device that could perform arithmetic operations. He later wrote that at the demonstration Hooke had taken close interest, even removing the back plate that covered the mechanism's inner workings. As Leibniz complained, this was enough for a man who was 'clever and mechanically-minded' to understand the principles on which his machine operated and make one of his own.[13]

Questions of trust and trustworthiness were at the very heart of natural philosophy in the seventeenth century, just as they are vitally important to scientists today in terms of both scientific research itself and its perception by the public. Hooke and his colleagues had to establish what sorts of mechanisms or structures for verifying data would be most appropriate for the new philosophy. In terms of interpersonal relationships, one theory has suggested that philosophers worked with each other following a code of gentlemanly behaviour, and that this enabled them to trust each other and work together in harmony. It was understood at the time that gentlemen did not lie (and therefore would not present false results to their colleagues), were disinterested (that is, they had no need to benefit financially from their work, unlike merchants and tradesmen) and subscribed to a code of politeness, thus preventing disputes that might threaten the cohesiveness of the philosophical community.[14] In this conception of the early Royal Society's social relations, Robert Boyle was the archetypal fellow: the independently wealthy son of the earl of Cork, he lived a pious and seemingly somewhat ascetic life, and never involved himself in acrimonious scientific disputes (illus. 10). When compared with someone like Boyle, Hooke's status is much less clear: though still a gentleman, he was the son of a minor clergyman, he worked for his living and he was not afraid to tell his colleagues that they were wrong. Was Hooke seen as a mere 'mechanic' by his wealthier colleagues?

There is some evidence to suggest that this was the case. After his visit to the Royal Society Leibniz wrote a worried letter to Oldenburg saying that he hoped Hooke would not 'incur the suspicion of intellectual dishonesty and want of true magnanimity' by copying his calculating machine; but Leibniz obviously thought Hooke was capable of doing exactly that. On a different occasion Oldenburg wrote to the Dutch mathematician and inventor Christiaan Huygens begging him not to be offended by Hooke's accusation that Huygens had stolen one of his ideas, adding: 'There are people who, not having seen much of the world, do not know how to observe that decorum which is necessary among honest folk.'[15] Flamsteed, smarting from a coffee-house encounter in which Hooke had contradicted his answer to a question about optics, wrote to a friend

that Hooke displayed 'the impudence of [a] solely Mechanick Artist'.[16] These instances, however, come from people who had had a bruising confrontation with Hooke: by casting him as a social, and therefore intellectual, inferior, they reinforced their own authority.

Hooke himself recorded very few instances in which he felt himself to have been socially slighted by colleagues, and his occasional grumbles in the privacy of his diary relate to specific points

10 Johann Kerseboom, *Robert Boyle*, 1689, oil on canvas.

of tension. For example, in 1677 Hooke and Nehemiah Grew were elected as the two Royal Society secretaries, but in Hooke's case the promotion was accompanied by some anxieties. After a meeting following his election, he noted in his diary: '[I] read notes [the minutes of the previous meeting] Distinctly. Grew placed at table to take notes. It seemd as if they would haue me still curator, Grew secretary. I stayd not at the Crown. I huffd at Hill at Jonathans, Sir J Hoskins being praesent.' Hooke believed that Abraham Hill was one of those who would have preferred Grew as secretary. Instead of going to the Crown Tavern for the customary post-meeting drink with his Royal Society colleagues, Hooke had retreated to Jonathan's. Perhaps Hoskins took Hill's side when Hooke 'huffd' at him, since the following day Hooke felt that he was 'much abated in his kindnesse', and even two weeks later, 'not really my freind'. Yet Hooke's concern that a faction in the society wanted to prevent him from acting as secretary surely had its roots in a longer-running set of problems that had resulted in Hooke losing faith in the society's officials. In 1675 Huygens had written to the society with the news that he had invented a watch regulated by a spiral balance, but Hooke insisted that he had in fact invented a spring-regulated watch a long time prior, in around 1658. He blamed Henry Oldenburg for revealing details of his early work to Huygens and was disappointed and hurt when the society's council took Oldenburg's part in the ensuing dispute.[17]

Hooke's own philosophical method does not provide much guidance about the social aspects of creating new knowledge. It is a manual very much focused on instructing readers how to regulate their own research, and while it encourages collaborations with other researchers, it does not explain how to manage these relationships or even hint that they might present any difficulties. One problem he did acknowledge was the difficulty of maintaining an impartial standpoint when considering the arguments of other philosophers, or more perniciously, one's own arguments. He argued that this had been a stumbling-block for philosophy through the ages: ancient philosophers, he claimed, 'were very supercilious, and very angry to be contradicted, and maintained their Opinions more because they had asserted them, than because they were true, they studied more to gain Applause and make themselves admired . . . than to perfect their Knowledge'.[18]

Hooke felt that this was a continuing problem for science, interfering with everyone's power to think logically and objectively about new ideas. On the one hand, men have an innate desire for power over the minds of others: to 'captivate and inslave Mens Minds to a Reverence or good Opinion of their Abilities, and Doctrines'. And for listeners, it is particularly difficult to avoid being taken in by these doctrines if we admire and respect the person from whom we have heard them. Hooke's suggested remedy for this problem is superficially straightforward: 'not to consider so much what the Person is that instructs, as how true the things are he asserts'.[19] That is, Hooke suggests that social or intellectual status should not be a factor in assessing the worth of an idea. Admittedly, he is discussing the problem of giving too much weight to the opinions of respected writers or teachers, rather than the opposite problem of dismissing contributions made by those with a lower social status. His whole method, however, focuses on scrutinizing and testing ideas and information before admitting them to be what his contemporaries referred to as 'matters of fact'. While many of his Royal Society colleagues would have believed that a gentleman's word should be trusted, and some would perhaps have extended this trust to experimental data, Hooke advocated a much more pragmatic approach: trust no one, especially not yourself. In practice, of course, everything was much messier.

FOUR

These My Poor Labours

n late July 1678 Hooke called at Moses Pitt's bookshop at the sign of the Angel in St Paul's Churchyard to borrow a book. He had seen the *Description of Spitsberg* previously, and he had mentioned it to his Royal Society colleagues during their recent conversation about what whales and other fish eat.[1] The *Description*'s author, Friedrich Martens, was a German naturalist, and his book described observations he had made while travelling on a whaling ship to Svalbard in the Arctic (illus. 11). Hooke wanted to borrow the book so that he could have it translated into English in order to read it aloud at a Royal Society meeting. At the same time, he returned to Pitt three more books: a celestial atlas, a book about the duties of those named as executors to wills and an act of parliament describing a new tax to raise money for a war against France. Presumably he also returned Martens's book after having it translated, since it does not appear in the catalogue of his library.

Books were Hooke's passion and, aside from coffee-drinking, his only indulgence. He browsed London's bookshops at every opportunity, in his infrequent leisure time and on his way to and from other appointments. Moses Pitt's was not the only bookshop that Hooke treated more like a lending library than a shop. He frequently borrowed and returned books, or bought and then returned them, sometimes in lieu of payment for another

volume. Either London's booksellers were relaxed about this habit among their clientele or Hooke was a favoured customer. He seems to have been on excellent terms with many of London's prominent booksellers, such as the Royal Society's official publisher, John Martyn. He also patronized more specialist vendors such as Everard Behagel and Jean Cailloüé, French booksellers on the Strand who sometimes imported specific titles at Hooke's request.

Books were Hooke's one extravagance, but he was scarcely alone in his addiction to print. Shortly after its foundation the Royal Society fellows had chosen as their motto 'nullius in verba', a quote from the Latin poet Horace that roughly translates as 'on no one's authority'. Seemingly, the motto signalled a rejection of the classical authors in whom earlier scholars and philosophers had placed so much faith: the new scientists

11 Whales illustrated in Friedrich Martens, *Spitzbergische oder Groenlandische Reise Beschreibung gethan im Jahr 1671* (1675).

would make their own authority, not take anyone else's word for it. In reality, however, the institutional and private records of early experimental philosophers show how important books, booksellers and printing houses continued to be, even at a time when natural philosophy seemed to prioritize direct experience gained from experiment and observation rather than reading. In fact, Hooke and his colleagues read omnivorously, across all branches of learning: Hooke's own library had sections on theology, history, classical literature, travel, languages and law, alongside medical, mathematical and other scientific works. The very first of the accomplishments that Hooke argued were absolutely necessary for a new philosopher was familiarity with the thinking of previous philosophers: 'their several Hypotheses, Suppositions, Collections, Observations, &c, their various ways of Ratiocinations and Proceedings, the several Failings and Defects both in their way of Raising, and in their way of managing their several Theories'. Hooke's point here was not that these older works would contain useful ideas or theories. His argument was that the act of reading and understanding philosophical works would help to prepare the mind for experimental philosophy by training it to think in a particular way:

> 'tis with Exercises of the Mind as with the Operations of the Body; one that has been bred up, and well skill'd in any Trade, shall go much more readily and handily about it, and make a much better piece of Work of a quite new Design in that Trade, than one that has not been at all us'd to such kind of Operations.[2]

Hooke's underlying assumption here – that doing natural philosophy is much like performing a skilled manual trade – is entirely characteristic. The only difference was that training in philosophy required a wide programme of reading rather than apprenticeship to a master craftsman.

Once this familiarity with existing knowledge and hypotheses was established, Hooke's method required the natural philosopher to build up what he called a 'Philosophical History', a storehouse of knowledge organized into a convenient order, ready for when the intellect might need it. This philosophical history was, in effect, a universal notebook:

> a brief and plain Account of a great Store of choice and significant Natural and Artificial Operations, Actions and Effects, ranged in a convenient Order, and interwoven here and there with some short Hints of Accidental Remarks or Theories, of corresponding or disagreeing received Opinions, of Doubts and Queries and the like.[3]

The material in this philosophical history would provide the 'Foundation Stones' on which the entire new philosophical structure would be raised. Not everything should be recorded, however. The philosophical historian should not be swayed by authority but should only repeat things 'affirmed by an inquisitive, judicious, and most strictly veracious Person'; writers who chose their words cautiously and never overstated their assertions. Hooke suggested adding a C, P or D in the margin next to citations, 'according as the Authority is Certain, Probable, or Doubtful', and he felt that there was no need to cite the authority's name, or at most the bare name and nothing else. The facts were what was important, not the person producing them. 'For', he wrote, ''tis not Epithets taken from Antiquity or Novelty, or Honour, or Greatness, or Witt, or Eloquence, or any other Learning but Experimental, that will be significantly added upon this Account.' Social station, eloquent rhetoric and so on were 'Visards', or masks: the new philosopher should disregard them and 'see only what Truth or Probability at least he can spy underneath'.[4]

We should keep in mind that Hooke was describing here his method for compiling a personal notebook for recording ideas rather than a publication intended for a wider audience. In his notebooks, the natural philosopher had to try to rid himself of any prejudice either for or against his material and simply see the observations, experiments or circumstances as they appeared. In Hooke's idealized programme, recording data in a structured and accessible format was the first step towards processing the information, weeding out anything not relevant to the question at hand and finally formulating a hypothesis. Sadly, we do not have any examples of such notebooks, so it is difficult to determine how closely Hooke followed his own recommendations. His existing manuscripts show that he was a habitual note-taker, often on random sheets of paper. It is possible that he compiled these notes into philosophical histories but that his posthumous editors later decided they were of no interest and discarded them, as they apparently did other ephemeral writings.

Hooke's method concentrates on how an experimental philosopher should create and record knowledge, but an equally important question for the new philosophers was how to communicate their findings. Should they use existing literary forms and language, or should they create something new? Francis Bacon had experimented with the essay, one of the first Englishmen to publish in this short, punchy format that made no claims to completeness but instead prompted its readers to continue the work begun by Bacon. Hooke also contributed to the contemporary discussion about a style of writing, or publishing, appropriate for science. He left a clue about his own (very Baconian) intentions in the preface to the first of his Cutlerian lectures to be published. He explained that he would not be following any particular order in his publications, merely providing

'Essays or Attempts' on subjects that may not have any connection with each other. He argued that this was the best way to fulfil the terms of Cutler's lectures, rather than writing an 'exact and compleat History' of just one subject. Instead, he would leave it to his readers to build on the foundation he was providing. Indeed, he argued that it would be better if more writers followed this method: rather than tormenting their readers with the 'nauseous Repetitions, and frivolous Apologies' required in the writing of longer books, they could instead 'inrich the Storehouse of Art and Nature with choice and excellent Seed, freed from the Chaff and Dross that do otherwise bury and corrupt it'.[5] The treatise that followed this introduction demonstrated his method: he stuck very much to his original point, explaining his apparatus at length (important so that others could trust his observations and even replicate them if they could set up similar equipment), providing his observational data and using the data to support his original argument.

This was the mode of writing that Hooke followed in all his published works intended for a philosophical audience. It reflected the fact that his research agenda was partly dictated by his lecturing duties: in the same introduction he explained that he would publish in turn on nature and art, the two topics of the Cutlerian lectures. Hooke argued, however, that 'Observations of Nature' and 'the Improvement of Art' were necessarily intertwined, and that it was only by bringing them together that knowledge would be advanced. We can see here one of the key points Hooke made in his philosophical method: the need for philosophers to investigate both natural and artificial objects.

Yet Hooke is more famous for a different kind of writing. His longest work, *Micrographia*, was aimed at a much wider audience than the members of the philosophical community, which means that we can also claim Hooke as a pioneer in the field of science communication. His philosophical method stipulated that a natural philosopher must communicate his or her findings, like Columbus, 'freely and impartially', with particular attention to the 'great Promoters and Benefactors' of the work.[6] This last point was no doubt tied to the necessity Hooke felt himself to be under to raise funds for his scientific projects, but he duly complied with his own advice and dedicated what he called 'these my poor Labours' to both Charles II and the Royal Society. We should not take this humility too seriously: he later punned that with his publication of *Micrographia* he was casting his 'Mite, into the vast Treasury of a Philosophical History'.[7] The allusion to the poor widow who had put her whole wealth of two small coins, or mites, into the church treasury would have been immediately intelligible to his biblically literate audience, who would also have appreciated Hooke's joke: one of his chapters described 'the wandering Mite'.

It was his publication of *Micrographia* in 1665 that cemented Hooke's reputation in England and brought him to the notice of continental philosophers. *Micrographia* was the world's first fully illustrated book of microscopy. It was beautiful and attention-grabbing, even unearthly, showing its readers a new world of previously unimagined tiny structures and mechanisms (illus. 12). The engraved plates were not simply there to accompany the text, they were works of art in themselves. The largest plates folded out from the book, expanding to four times their original size, so that one astonished reader wrote to a friend that the book contained a flea and a louse 'as big as a cat' (illus. 13).[8] Throughout the early 1660s Hooke had presented new microscopical observations at Royal Society meetings: moss one week, the pores in a section of cork the following week and 'leeches in vinegar' (vinegar eels, nematodes found living in raw vinegar) the week after that. The fellows were clearly fascinated and often requested that Hooke look at specific things: a sample of petrified wood; sage leaves; 'viper powder' (that is, adders

dried and ground into powder); and rainwater containing 'a great number of little insects'.[9]

When he came to write about his observations, Hooke's own sense of fascination and wonder at the tiny mechanisms he found in nature came through particularly strongly. He started by describing a few man-made objects: the tip of a needle, the edge of a razor and a printed full stop (illus. 14). Partly intended as an elaborate joke about geometry (Euclid began his geometry with definitions of the point and the straight line), the main point of these initial observations was to completely overturn his audience's beliefs about the everyday material world. Hooke demonstrated that rather than being sharp and defined, the point of the needle was blunt and rounded; the razor's edge was pitted and scarred; and the full stop ragged and ugly like a 'great splatch of London dirt'.[10] One of Hooke's fundamental points in *Micrographia* was that on a microscopic level beauty and precision could only be found in the natural world: the closer one looked, the more perfect these natural structures appeared.

12 Hooke's illustration of mould growing on the leather cover of one of his books, printed in *Micrographia* (1665).

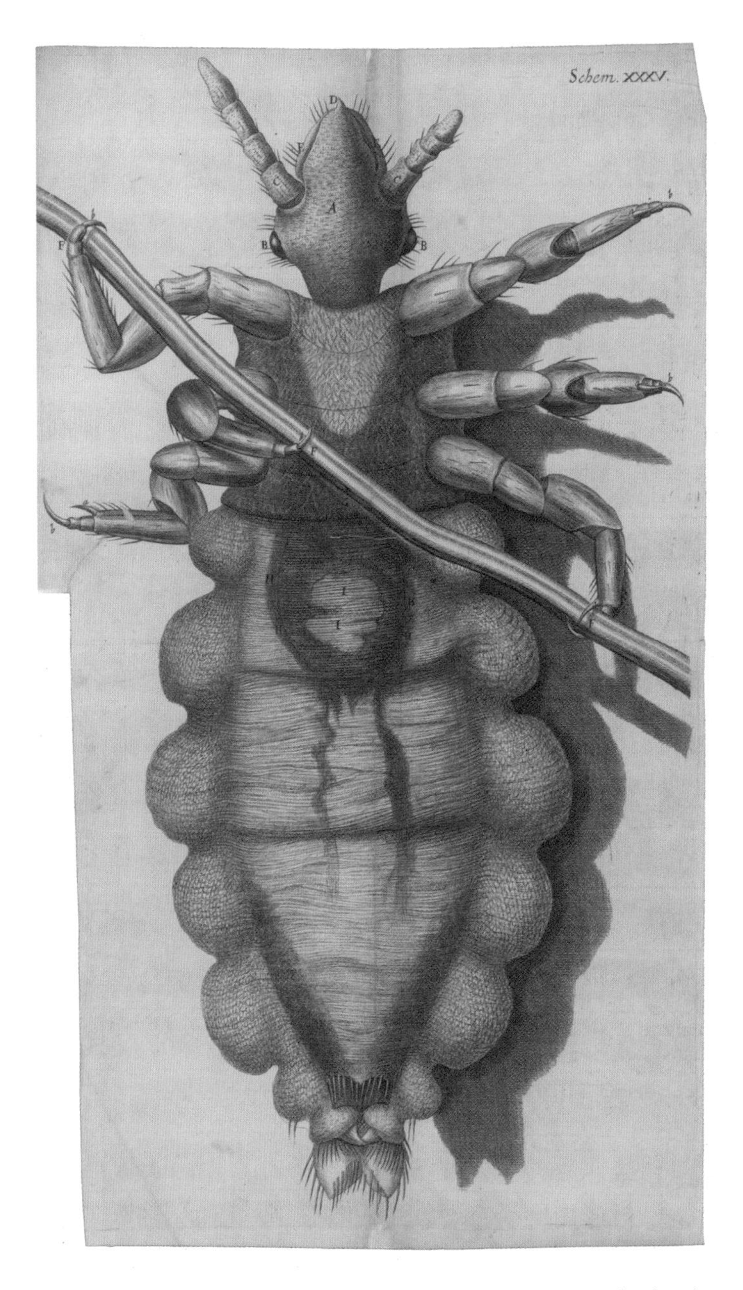

13 Illustration of a louse holding a human hair, printed in *Micrographia* (1665).

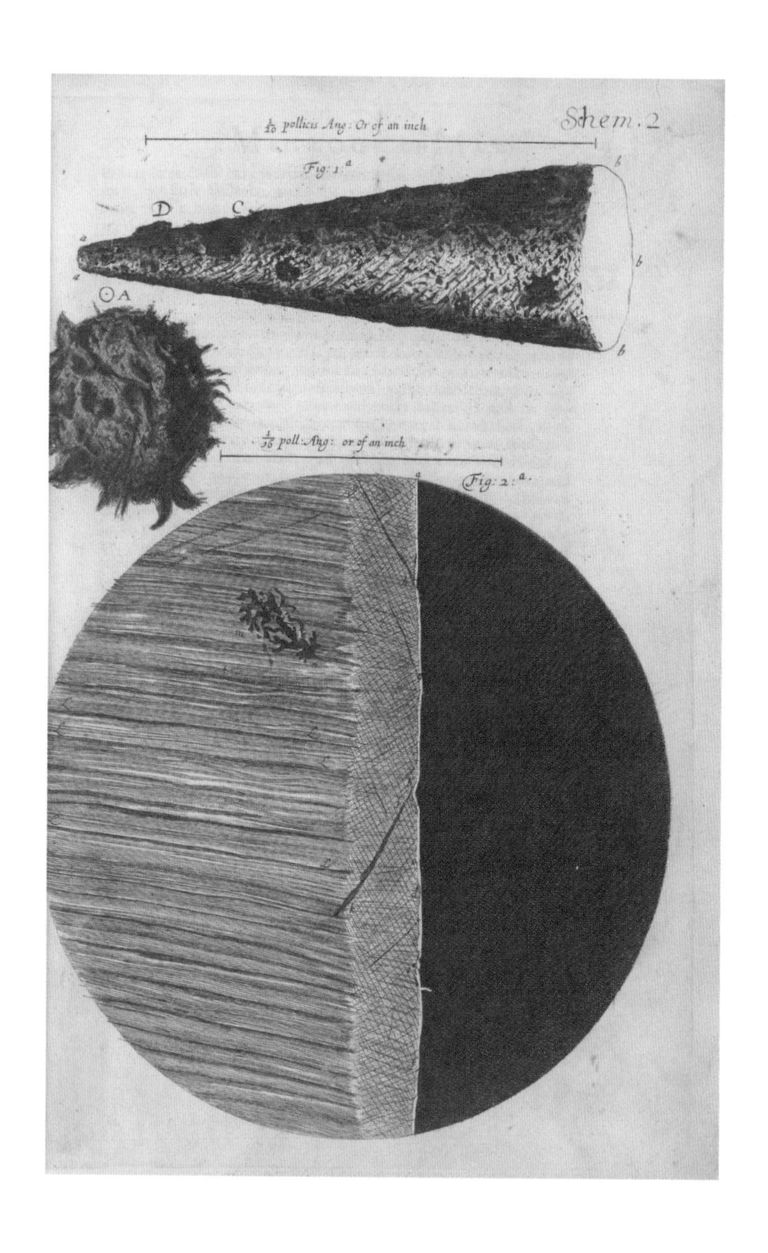

14 Illustration of a printed full-stop, the tip of a needle and the edge of a razor, printed in *Micrographia* (1665).

Hooke's language throughout *Micrographia* was vivid and appealing: he made jokes, quoted other authors and used rhetoric to bring his subjects to life. His flea wore 'a curiously polish'd suit of sable Armour'; he joked that the head-louse was like an impudent courtier or social-climber, liking 'nothing so much as a Crown', 'troubled at nothing so much as at a man that scratches his head, as knowing that man is plotting and contriving some mischief against it'. Hooke himself was a constant presence in his text, explaining to the readers how his observations were made and encouraging them to try microscopy for themselves. The diarist (and Hooke admirer) Samuel Pepys famously sat up all night reading his copy of the book, calling it 'the most ingenious book that ever I read in my life'.[11] Hooke's writing style in *Micrographia* was aimed at a wide audience who were mostly either unaware of the activities of the experimental philosophers or sceptical about the benefit of experimental natural philosophy. His philosophical method, probably composed around the time of *Micrographia*'s publication, stipulated that new philosophers should register their work 'in the plainest, shortest, and most significant Description' possible. This followed the line taken by his Royal Society colleagues, who insisted that philosophers should reject flowery language, though they often ignored their own precepts. But Hooke's own lively style in *Micrographia* was linked with his underlying aim, which was to open his readers' eyes to the 'new visible World' revealed by the microscope, and to encourage them to make observations of their own.[12]

Hooke's research in *Micrographia* continued to influence his thinking, and the direction of his research, throughout his life. In *Micrographia* he first described his theory that light travelled in waves, part of a long discourse on colour sparked by his observation of coloured rings in mica, or 'Muscovy Glass'.[13] He observed hollow structures in a very thin slice of cork, naming them 'cells' because they looked like the cells in a honeycomb.

He suggested that 'figured stones' found in the shape of shellfish were in fact the petrified remains of actual shells, and not the strange productions of the earth made in imitation of seashells. He also took the opportunity to restate and elaborate on the theory of matter that he had originally put forward in 1661, in his first surviving published work. This was the idea that some liquids readily mix with other liquids or adhere to solids and others do not, properties that Hooke designated as 'congruity' and 'incongruity'. *Micrographia* also influenced his working methods. His observations of everyday materials and specimens that came to hand demonstrated to him the fact that advances in knowledge could come just as easily from observing prosaic objects – pieces of cloth or wood, snowflakes, fleas and lice – as exotic specimens, a point that he reiterated frequently in his scientific method.

Micrographia was so popular that a second edition was printed in 1667. With these major exceptions, most of Hooke's publications appeared in the 1670s. His earliest printed work seems now to be lost. His next publication was a short treatise with a long title, *An Attempt for the Explication of the Phaenomena Observable in an Experiment Published by the Honourable Robert Boyle, Esq; in the* XXXV *Experiment of His Epistolical Discourse Touching the Aire. In Confirmation of a Former Conjecture Made by R. H.* Published in 1661, this short work discussed experiments Boyle and Hooke had made on the capillary action of liquids in slender glass tubes. As the title makes clear, Hooke initially positioned himself as Boyle's assistant: it was Boyle's name that appeared in full in the title, and Boyle's experiment on which Hooke was commenting. It is possible that Boyle had also unconsciously influenced Hooke's writing style, at least in the title. Boyle's own style was regarded even by his contemporaries as somewhat long-winded, so much so that one contemporary satirist parodied Boyle's language for comic effect in a burlesque piece attacking the Royal Society.

It was over a decade later that Hooke published the first of his Cutlerian lectures to be printed. It appeared in 1674 under the title *An Attempt to Prove the Motion of the Earth from Observations Made by Robert Hooke Fellow of the Royal Society*. In this short treatise Hooke presented evidence for the Copernican heliocentric system supplied by his own observations using the zenith telescope he had installed in his rooms at Gresham College. This telescope observed the sky directly above it. As Hooke explained, and indeed depicted in a diagram showing his apparatus (illus. 15), he had 'opened a passage of about a foot [30 cm] square through the roof of my lodgings' in order to accommodate the telescope, which had no tube, since that would have been extremely inconvenient given that it would have had to pass through two rooms. Seemingly the feelings of the Gresham College trustees about the inconvenience of having a hole cut in their roof were less important to Hooke. Later they agreed to the construction of a purpose-built 'turret' for astronomical observations at the top of Hooke's rooms.

In his *Attempt* Hooke also set out three 'suppositions' that he claimed would be the foundation for a new understanding of the universe. First, that all celestial bodies have 'an attraction or gravitating power towards their own Centers' and that they attract all the other celestial bodies within the sphere of their activity. Second, that all bodies 'put into a direct and simple motion' will continue to travel in a straight line until acted upon by another power that deflects them into a 'Circle, Ellipsis, or some other more compound Curve Line'. Third, 'that these attractive powers are so much the more powerful in operating, by how much the nearer the body wrought upon is to their own Centers'. He admitted that he had not done the experiments to demonstrate exactly how these powers worked, but he hoped that someone could be found to do it, promising that person that 'he will find all the great Motions of the World to be influenced

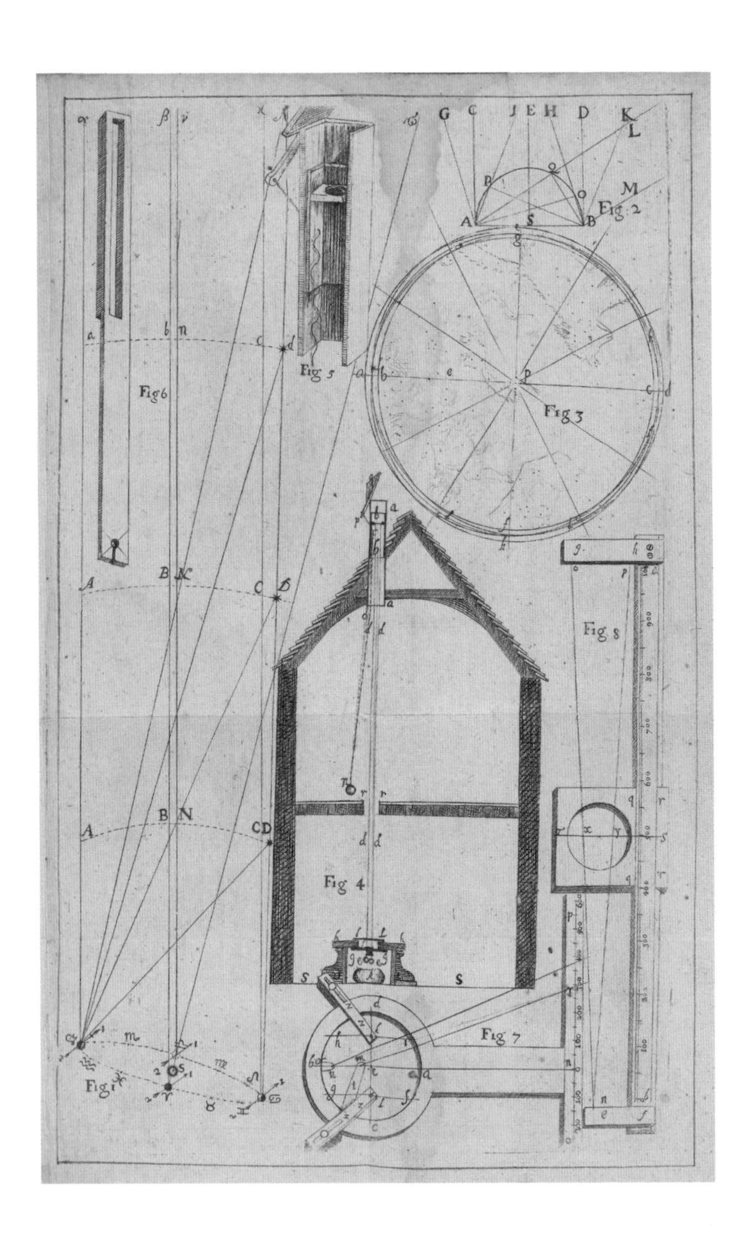

15 Diagram (fig. 4) showing Hooke's zenith telescope installed in his rooms at Gresham College, 1674. The object glass was fixed in a moveable frame at the top and the eye glass was in a 'slider' fixed in a table on the ground floor,

by this Principle, and that the true understanding thereof will be the true perfection of Astronomy'.[14] Hooke revisited these ideas in correspondence with Isaac Newton in 1679, prompting Newton to revise his model of planetary motion to take into account Hooke's theory of gravitation.

In this period Hooke was preoccupied by the problem of making accurate astronomical observations, and his second publication in 1674 addressed this same subject. In December that year he published *Animadversions on the First Part of the Machina coelestis of the Honourable, Learned, and Deservedly Famous Astronomer Johannes Hevelius Consul of Dantzick; Together with an Explication of Some Instruments Made by Robert Hooke, Professor of Geometry in Gresham College, and Fellow of the Royal Society.* This was a much longer volume than the *Attempt to Prove the Motion of the Earth*. The *Animadversions* had been prompted by the work of the Polish astronomer Johannes Hevelius, whose instruments and observatory at his house in Danzig (modern Gdańsk) were famous throughout the European astronomical community. Hevelius used sextants and quadrants with open sights, adjusting the instrument and then using the naked eye to measure the position of a star in the night sky (illus. 16). Hooke argued that however large and accurate Hevelius's instruments might be, it was impossible for the human eye to distinguish very small angles. Hooke suggested that Hevelius use telescopic sights, which he argued allowed for much more precise measurements. Though he stressed his respect for Hevelius as a careful observer, Hooke completely undermined Hevelius's method, and thus his observational data. Hevelius was justifiably upset, but what the *Animadversions* really demonstrates is Hooke's passion for instruments, rather than for philosophical dispute. His method rested on the belief that the human senses need the help of scientific instruments in order to create new knowledge. At the same time, he was committed to developing

about 12 metres (40 ft) below. The lid at the top could be raised or lowered using a string, to protect against rain.

16 Hevelius and his wife Elisabeth using a sextant, illustration from Johannes Hevelius, *Machina coelestis pars prior* (1673).

new mechanical devices that would improve daily life for his contemporaries.

This theme continued to inspire Hooke's publishing programme in the mid-1670s. In 1675 he published *A Description of Helioscopes, and Some Other Instruments*, and in 1676 *Lampas: or, Descriptions of Some Mechanical Improvements of Lamps and Waterpoises. Together with Some Other Physical and Mechanical Discoveries*. These treatises focused on instruments and apparatus invented by Hooke to solve specific problems (for example, how to observe the Sun safely, or how to make a better lamp), and discussed related discoveries or hypotheses. In March 1678 he published *Lectures and Collections Made by Robert Hooke*, a work made up of two parts, 'Cometa' – dealing with observations of a recent comet, and other astronomical work – and 'Microscopium', containing work on microscopy and anatomy. In November of the same year, Hooke published *Lectures de potentia restitutiva; or, Of Spring, Explaining the Power of Springing Bodies*, to which was added 'some collections' of work by other people. These two works, *Lectures and Collections* and *Lectures of Spring*, were a different form of publication. In these two books, Hooke's own work on comets, microscopes and the physics of springs acted as a focus for associated observations and conjectures written by other experimental philosophers in letters and papers that had been sent to Hooke. Thus the first part of Hooke's *Lectures of Spring* explained his theory of elasticity, now known as Hooke's law, which he initially stated in Latin and then translated as 'The Power of any Spring is in the same proportion with the Tension thereof.' This led into a discussion of the underlying physics of springiness, which Hooke again argued stemmed from the principles of congruity and incongruity. But alongside his own discussion he printed the French physicist and mathematician Denis Papin's description of a new design for a pump and several accounts by different authors of natural springs and fountains, and further natural phenomena such as caves, earthquakes and volcanoes, all of which he believed were in some way connected with the topic of spring. The structure of the volume reflects Hooke's insistence on the interconnection between natural and artificial phenomena, and the importance of combining knowledge of both in any investigation into physics.

Hooke's move away from the single-authored treatise as the preferred form of scientific publication may have partly been prompted by his appointment as one of the Royal Society's secretaries in 1677. He had actively campaigned for this role after the death of the society's first secretary, Henry Oldenburg, earlier in the year. Oldenburg had been Hooke's equal in terms of the significance of his contributions to the society, managing all the correspondence and instigating many connections with philosophers abroad, taking minutes at the meetings and starting his own publishing programme. His death left a huge gap in the society's operations, and the council decided to elect two men to replace him. As we have seen, Hooke had come into conflict with Oldenburg a couple of years earlier, suspecting him of passing details of his spring-regulated watch design to Christiaan Huygens. He also believed Oldenburg had been deliberately omitting some of his contributions to Royal Society meetings from the minutes. By comparing his own notes with the society's journal books he did find several occasions where this was the case, something he was quick to rectify when he began taking the minutes himself after Oldenburg's death.

Oldenburg's most significant contribution to the society had been his publication venture, the scientific journal *Philosophical Transactions*. Oldenburg first printed *Philosophical Transactions* in 1665, and over the next twelve years its monthly instalments became required reading among philosophical men across Europe. The journal contained short articles, letters and book reviews on scientific topics, mostly material that had been sent to the Royal Society or read at one of their meetings. Despite initially being Oldenburg's project rather than an official institutional undertaking, the fellows appreciated the publicity and the opportunity to demonstrate priority by publicizing new inventions or theories. Publication lapsed after Oldenburg's death, although the society's council made some half-hearted attempts to encourage Hooke to take over as editor. Then, in 1679, Hooke began to publish his own periodical, the *Philosophical Collections*. The contents were very similar to the articles in the *Philosophical Transactions*, although perhaps skewed somewhat towards Hooke's own interests and the contributions of his friends and associates. Hooke occasionally published his own material in the *Collections*, such as an 'Optical Discourse' that had been presented at a meeting of the Royal Society 'proposing a way of helping short-sighted or purblind Eyes', and a treatise about the best design for a mill with horizontal vanes.[15] Other articles were translations from continental publications, or notices about new books that had been published. Perhaps it proved to be a more time-consuming task than Hooke would have liked, because only seven volumes appeared at rather irregular intervals between October 1679 and April 1682. After this the *Collections* lapsed, and the Royal Society revived the *Philosophical Transactions* under a new editor in 1683.

Apart from some new editions of earlier works, and the occasional *Philosophical Transactions* article, these were the last of Hooke's publications. It is not clear why Hooke chose not to publish further lectures of his own after this burst of activity in the 1670s. In his biography Waller claimed that Hooke had become more reserved in his later years and had 'seldom left any full Account' of his Cutlerian lectures, though he had intended to revise them for the press.[16] He did, however, take an active interest in publishing other writings. One such volume was a description of Ceylon written by Captain Robert Knox, a merchant sailor who had been held captive on the island for more than nineteen years. Knox had finally escaped and returned to England in 1680, much to the interest of Hooke, who was already acquainted with Knox's brother. Knox's narrative, published as *An Historical Relation of the Island Ceylon* (1681), is part anthropological account, part adventure story, telling of Knox's capture, his life in Ceylon and his ultimate escape. Hooke wrote a very supportive preface for the volume in which he playfully assured readers that whatever their interests they would find themselves 'taken Captive' by the narrative. He also made the serious point that the philosophical world needed more trustworthy accounts of foreign parts, like this one. It was unfortunate that even relatively recent books about places such as the West Indies fell far short of the knowledge 'which may be obtained from divers knowing Planters [colonizers] now residing in London'.[17] Hooke suggested that this was partly because travellers were not sure what they should record, and partly due to a lack of 'fit Persons' to act as promoters and ghost-writers for travel accounts. In fact, the Royal Society had produced printed lists of questions for travellers in the 1660s, requesting detailed information about everything from local weather conditions, geography and flora and fauna to the customs of indigenous people. It is interesting that Hooke also highlighted the role of those who were not travellers themselves, but who conversed with and questioned those who had returned from foreign parts and helped to compile the answering information into a history of the kind that Knox had produced. There is no evidence that Hooke acted as a mediator in this way for Knox, but one cannot help but hear an echo of the coffee-house conversations in which he plied travellers with questions about everything from what it was like to experience a volcanic eruption to the best cure for a rattlesnake bite.

Books were also a way to signal admiration or to cement friendships. Hooke always recorded lists of the friends and colleagues to whom he gave presentation copies of his own work. There was a fine-grained hierarchy here, with patrons such as Robert Boyle, Sir John Cutler, Hooke's former schoolmaster Richard Busby and Royal Society president Sir Joseph Williamson receiving gilt-edged copies, sometimes with marbled endpapers, and other friends and coffee-house associates receiving ordinary copies. In 1693 the Somerset natural philosopher John Beaumont visited Hooke to present him with a copy of his new book. Hooke later wrote in his diary, 'when he was Gone I found he had Dedicated it to me'.[18] Beaumont's book was a response to a previous publication by another natural philosopher, Thomas Burnet, describing his elaborate theory of the formation of the earth, a topic on which Hooke had lectured at the Royal Society. Beaumont had read the first part of his manuscript to Hooke and others at Jonathan's coffee-house a few months prior, and this dedication was a fitting tribute from a younger philosopher to an acknowledged leader in the field.

Despite his interest in colleagues' publications and his love of reading, Hooke did not consider a book-length treatise to be absolutely fundamental to the progress of science. In the nineteenth and early twentieth centuries historians of science habitually criticized Hooke, when they noticed him at all, for working across too many fields rather than single-mindedly pursuing one subject. They characterized him as a dilettante who never brought his ideas to fruition, or more charitably, suggested he simply did not have time to work steadily on a single problem. Of course, the model scientist they had in mind was Isaac Newton, whose books *Philosophiae naturalis principia mathematica* (The Mathematical Principles of Natural Philosophy, 1687) and *Opticks* (1704) formed a lasting monument to his genius. In contrast, Hooke's publications looked a little scanty. And yet, his publishing programme fitted his own scientific method. He was one of the first to make sustained use of what would go on to become the primary form of scholarly publication, the scientific journal, and he believed that in doing so he was creating the conditions for future advances in knowledge.

FIVE

A Man Who Is Mechanically Minded

Brushing against the leaf of a stinging nettle causes pain, but in the seventeenth century, no one knew why. For such a common experience, Hooke thought this was remarkable. Setting out to investigate with his microscope, he found that the whole surface of the nettle plant was 'very thick set' with 'Bodkins' or 'sharp Needles' (illus. 17). These needles were hard, stiff and transparent, and importantly, they were hollow. Watching closely through a special microscope fashioned 'almost like a pair of spectacles', Hooke thrust the nettle leaf against his skin and watched a liquid rise through these needles from flexible 'baggs' at the base of each needle. This liquid descended again when he took his hand away. He guessed that the little bags at the base of the needles were full of a 'poisonous juice' produced by the plant, and the needles acted as tiny syringes, conveying the poison 'into the interior and sensible parts of the skin' where it burnt or corroded the flesh and caused the painful stinging sensation.[1]

Hooke was never content simply to understand how a mechanism – natural or artificial – worked; instead, he looked for ways to put his mechanical knowledge to use. After discovering the secret of nettle stings, he theorized that it was the poison's delivery under the skin that made it so potent, because it was immediately dissolved and incorporated into the body. He went on to suggest that doctors might be able to use this principle to benefit their patients: why not experiment further with using syringes to deliver medicine directly into the veins? In this way, 'several obstinate distempers of a humane body, such as the Gout, Dropsie, Stone, etc, might be mastered, and expelled'. A principle from the natural world could potentially improve human life.[2]

Hooke's enthusiasm for the tiny mechanisms he observed in nature was just a part of his enduring fascination with instruments and mechanical contrivances of all kinds. As we have seen, as a boy on the Isle of Wight he amused himself making 'little mechanical Toys', including a working clock made of wood, and a small model ship complete with rigging and guns.[3] At Westminster School, John Aubrey wrote, 'he was very mechanicall, and (amongst other things), he invented thirty severall wayes of Flying.'[4] His talent for immediately understanding

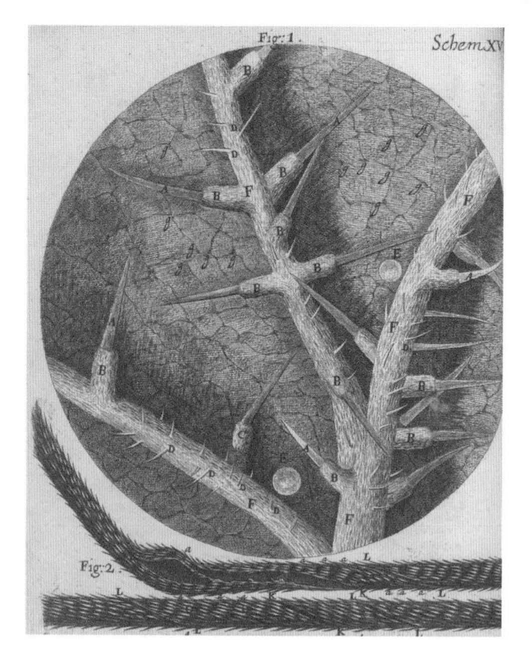

17 Microscopic detail of a stinging nettle leaf, printed in *Micrographia* (1665).

how a mechanism worked, and how it might be improved, was acknowledged by his friends and colleagues. Leibniz worried that the 'mechanically-minded' Hooke might be able to copy his calculating machine, and Flamsteed mocked him for being merely a 'Mechanick Artist'. Hooke's biographer and friend Richard Waller spoke of his 'Mechanick Genius', and recalled approvingly his intuitive grasp of mechanisms of all kinds: it had 'been often observed by several persons', he wrote, that whatever experimental apparatus Hooke demonstrated at the Royal Society, he always explained it without confusion, 'clearly and evidently'.[5] Aubrey, another close friend, called Hooke 'the greatest Mechanick this day in the World'.[6] Hooke devoted a large part of his experimental life to mechanical devices, but how did his 'mechanick genius' connect with his philosophy of science?

Part of Hooke's enthusiasm for instruments stemmed from his belief that through technology, mankind could overcome the deficiencies of the human senses. Although we rely on our senses to gather data about the natural world, they are limited by the constitution of the human body. But what more could be learnt if mechanical aids could be designed that would enlarge the senses? Telescopes and microscopes had already demonstrated the possibilities of assisting sight. Hooke argued that the other senses were equally capable of expansion. Perhaps, he admitted, the sense of hearing was not quite as useful as sight. And yet, by using an otocousticon, very distant sounds might be heard. Using this device could enable sailors to hear thunder earlier, so 'Ships at Sea might perhaps discover an Enemy of Weather coming . . . as well as they can now discern an Enemy's Ship by the Sight'. Or perhaps an instrument might be invented that would allow a doctor to hear what was going on inside a human body: to 'discover the Works perform'd in the several Offices and Shops of a Man's Body, and thereby discover what Instrument or Engine is out of order, what Works are going on at several Times, and lie still at others'.[7] He even raised the possibility of 'the hearing of Noises made as far off as the Planets', although he admitted that it seemed 'a very extravagant Conjecture', unlikely to be achieved. However, it was part of his philosophy to believe that with the right technology anything might be possible: dismissing out of hand the ideas that most people thought were 'mad, foolish and phantastick' would never advance knowledge.

Hooke was by no means the only experimental philosopher to dream of using science to repair or enhance the naturally limited human body. At some point in the mid-seventeenth century Robert Boyle made a list of things he hoped science would achieve. He included some perennial favourites such as 'The Art of Flying' and 'The practicable and certain way of finding Longitudes'. Other goals involved improvements to the human body. At the top of the list stood 'The Prolongation of Life', followed by 'The Recovery of Youth, or at least some of the Marks of it, as new Teeth, new Hair colour'd as in youth'. Further down the list came 'The Attaining Gigantick Dimensions', and 'The Emulating of Fish without Engines by Custome and Education only'.[8] While Boyle clearly did see the body as a site for scientific progress, his focus was on changes to the actual fabric of the body: eternal youth, larger stature and the ability to swim underwater for long periods enabled not by 'Engines' but through special training. These goals potentially intrigued Hooke, but for him, they may have seemed something of a distraction from what he saw as the true purpose of improvements to the human body: the ability to investigate the natural world in ever-closer detail.

Hooke's faith in the power of instruments and mechanical devices to enable wondrous new powers of perception rested on his belief that the universe was itself fundamentally mechanical. He, and others who subscribed to the 'mechanical philosophy' popularized in particular by Descartes, believed that the universe essentially consisted of small particles of matter, invisible to the naked eye, interacting with each other, and that these particles did so according to universal laws. For Hooke in particular, one outcome of this was that natural and man-made mechanisms operated according to the same principles. His work on *Micrographia* seemed to him to confirm this theory, and it was not just plants that operated like machines. Trying to explain why it was important for philosophers to examine a mixture of perfect and imperfect specimens, Hooke used the example of a man. As a man gets older, he wrote,

> the Parts grow dryer and stiffer, and less fit for Motion; the Natural Moisture is grown too thick and slug[gish], and the parts begin to shrink and shrivel . . . the Juice in many parts of the Body become[s] so charg'd with excrementitious parts thereof, that it turns into a kind of hard or stony Consistence: and like an old decayed and foul piece of Clock-work, here a Pivot is worn loose in his Socket, there the Oyl is thickened with Dust and Filth, as almost to stop the Motion.[9]

He was making the point that it is useful to compare old and new specimens, but his metaphor suggests that in some ways, Hooke saw even the human body as a mechanical device, just as prone to drying up and wearing out as an old clockwork that had not been oiled.

As might be expected, not everyone subscribed to a philosophy in which the human body, or the universe itself, seemed to be no more than a piece of clockwork. Hooke's most vigorous defence of the mechanical philosophy came as a response to the work of the philosopher and FRS Henry More. More had argued that everything happened in the natural world through the agency of a 'spirit of nature', or 'Hylarchick Spirit': an invisible power that caused the growth and life of living things, and worked to produce seemingly inexplicable physical phenomena such as magnetism and tides. Hooke not only found this theory unnecessary and preposterous, but he argued that it represented a danger to the advancement of knowledge. In the middle of his discussion of a way to keep the oil topped up at a steady level in a lamp, he digressed to attack More's spirit and to reaffirm his own belief that the natural world operates by mechanical laws. His diatribe shows how passionate he was in championing both his physics and his view of science (and it is worth quoting at length to get a sense of Hooke's style):

> We see then how needless it is to have recourse to an Hylarchick Spirit to perform all those things which are plainly and clearly performed by the common and known Rules of Mechanicks, which are easily to be understood and imagined, and are most obvious and clear to sense, and do not perplex our minds with unintelligible Idea's of things, which do no ways tend to knowledge and practice, but end in amazement and confusion.
>
> For supposing the Doctor [that is, Henry More] had proved there were such an Hylarchick Spirit, what were we the better or the wiser unless we also know how to rule and govern this Spirit? . . . This Principle therefore at best tends to nothing but the discouraging Industry from searching into, and finding out the true causes of the Phenomena of Nature: And incourages Ignorance and Superstition by perswading nothing more can be known, and that the Spirit will do what it pleases. For if all things be done by an Hylarchick Spirit, that is, I know not what, and to be found I know not when or where, and acts all things I know not how, what should I

> trouble my self to enquire into that which is never to be understood, and is beyond the reach of my Faculties to comprehend? Whereas on the other side, if I understand or am informed, that these Phenomena do proceed from the quantity of matter and motion, and that the regulating and ordering of them is clearly within the power and reach of mans Industry and Invention; I have incouragement to be stirring and active in this inquiry and scrutiny, as where I have to do with matter and motion that fall under the reach of my senses, and have no need of such Rarified Notions as do exceed Imagination and the plain deductions of Reason therefrom.[10]

Hooke's frustration here is palpable. It was not simply the fact that More's mysterious spirit was unnecessary as an explanation for nature's workings. For Hooke, the key problem was that ascribing natural functions to an unknowable and ungovernable force meant that humankind was doomed to remain in the torpor of ignorance that Hooke's philosophical method was designed to fight against. His whole programme was based on the profound sense of optimism that drove his own research: he believed unwaveringly that with time (and possibly with the assistance of his philosophical algebra), science would help humans to answer even the most complex questions about the natural world.

It is unsurprising, then, that Hooke's philosophical method stipulated that the ideal experimental philosopher must be skilled in mathematics and mechanics. While mathematics taught people how to think clearly and logically, mechanics demonstrated how the world worked. The knowledge gained through the study of mechanical devices could then be used to understand the inner workings of nature:

> For though the Operations of Nature are more secret and abstruse, and hid from our discerning, or discovering of them, than those more gross and obvious ones of Engines, yet it seems most probable, by the Effects and Circumstances; that most of them may be as capable of Demonstration and Reduction to a certain Rule, as the Operations of Mechanicks or Art.[11]

He warned his readers, however, that there was as yet no road map for pursuing this line of enquiry, and that the various 'Turnings and Windings' and other difficulties would need to be managed by the philosopher's own initiative and perseverance (just as Columbus had negotiated the obstacles in his own journey of exploration).

Hooke himself devoted much of his time to inventing and improving scientific instruments and mechanical devices of all kinds. His work in this field was so wide-ranging that it is difficult to describe in a brief overview. Very broadly, there were two overarching and interconnected themes. The first was driven by Hooke's desire to improve the human capacity to understand or measure the natural world: telescopes, microscopes, clocks and watches, barometers, helioscopes (an instrument used to observe the Sun), his equatorial quadrant, marine depth-sounding devices and so on. The second was driven by another fundamental belief of Hooke's: science could – and should – benefit humanity more directly. This rather miscellaneous category includes inventions such as flying machines (never fully explained), the self-regulating lamp, a machine for making bricks and an improved cider-mill, a windmill with horizontal vanes, improvements to chariots and saddles and new methods of printing and drawing. Other useful mechanisms were incorporated into these devices, such as the universal joint, part of the equatorial quadrant (illus. 18). But these divisions

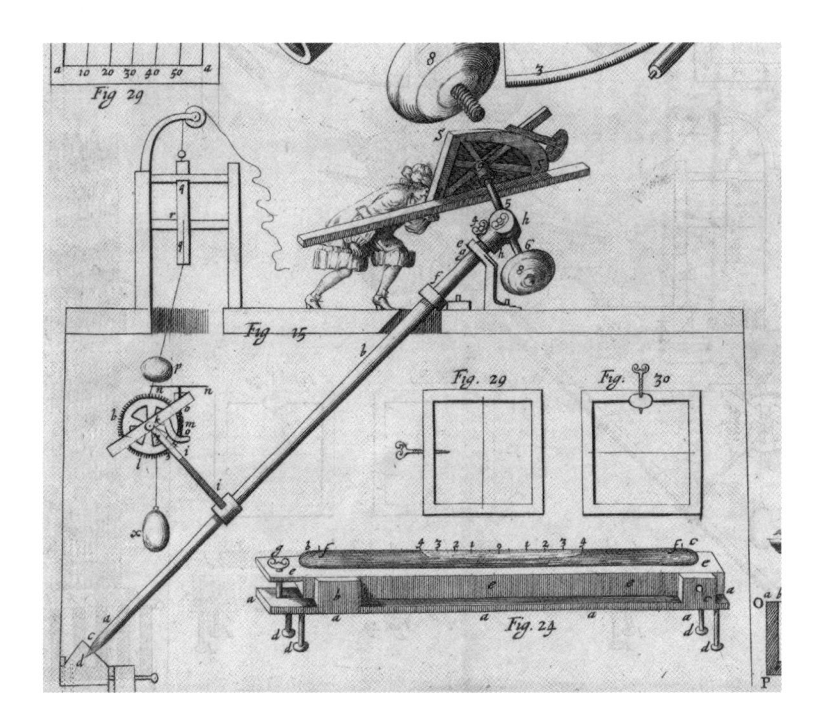

should only be seen as a way of trying to impose order on to Hooke's wildly productive output, and in reality many of his mechanical devices had elements of both categories. Where, for example, should we place his device for measuring and automatically recording the weather, his diving apparatus or the idea of a marine submersible 'so contrived as to row or be moved from within under the surface of the water'?[12] In his 1675 publication *A Description of Helioscopes, and Some Other Instruments*, Hooke promised his readers a description of a new 'Turning Engine', or lathe, that could be used

> to turn all manner of Conical Lines . . . all manner of Foliage and Flower-work, all variety of Basket or Breaded-work [braided work], all variety of Spiral and

18 Hooke's clockwork-driven equatorial quadrant, described in *Animadversions on the First Part of the Machina coelestis* (1674).

> Helical-work, serving for the imitation of the various forms and carvings of all sorts of Shells; for cylindrical and conical Screws; all variety of Embossments and Statues; all variety of edged and Wheel-like work; all variety of Regularly shaped Bodies.[13]

He never produced a detailed plan of this 'engine', but his description shows that, as always, Hooke was equally as aware of his audience in the community of artists and craftsmen, who might be interested in a quicker mechanical method of producing 'Foliage and Flower-work' or carvings of shells, as he was the instrument-makers who wanted to fabricate screws or other 'Regularly shaped Bodies'.

Hooke was fortunate to live in the centre of an industrious city. During Hooke's lifetime London hosted a growing number of instrument-makers, many of whom worked for the Royal Society. Prominent among them was Richard Reeve Sr, a skilled ivory-turner who in the 1650s and early 1660s made telescopes and microscopes for Hooke, Christopher Wren and other Royal Society fellows. One of his telescopes was set up in the courtyard at Gresham College for use by Hooke and others. After Reeve's death in 1666, Christopher Cock took over as the society's preferred supplier of optical instruments. He may have initially worked with Reeve, and like Reeve his business was located in Longacre. His microscopes and telescopes were so highly regarded that the Royal Society received orders from correspondents abroad. Hooke visited Cock (whom he knew as 'Cox') frequently throughout the 1670s, buying lenses, borrowing tools and keeping up to date with new techniques and materials. In his very first diary entry, in March 1672, Hooke wrote that he had 'told Cox how to make Reflex glasses by Silver and hinted to him Making them by printing'. He expanded on his 'hint' afterwards at a meeting of the Royal Society, where he suggested that it might be possible to make concave mirrors by using the mill at the Royal Mint to stamp out pieces of silver plate between two dies, one convex and one concave. In October 1677 he found Cock polishing a '100 foot glasse' for Sir Jonas Moore (a telescope lens with a focal length of 30 metres); in July 1679 he saw Cock's concave lenses, which had been polished in a new way; in 1681 he saw a 'strange Refracting White glasse cutt with Diam[ond] powder'. In September 1675 Cock shared with Hooke 'the way of polishing with white marble tooles couerd with thin paper. and Dusted with tripoly' (tripoli is a fine-grained earth that was used for polishing by goldsmiths and others). This was a continuation of a conversation Hooke had had a few days prior with Cock and 'the Ingenious fryer of Leige' about grinding and polishing lenses. The friar of Liège has not been identified, but his presence in Cock's workshop suggests that London's instrument-makers were a draw for visiting continental philosophers. Hooke sometimes watched Cock at work, recording in January 1676 that he 'Saw him polish an exellent 12 foot [3.6 m] glasse by changing place of the tool.' On this occasion Hooke stayed to smoke three pipes with Cock, bought 'a piece of white plate' and recorded Cock's promise to 'grind me a glasse of any shape if I would shew him my new way'.[14]

Hooke was not merely an interested bystander in Cock's workshop. He continued his own experiments into new methods of making lenses throughout the 1660s and '70s, part of his broader drive to improve telescopes and microscopes, an endeavour that occupied philosophers and instrument-makers across Europe. He had published a description of a lens-grinding 'Engine' in *Micrographia*, which he claimed could make lenses 'of any length or breadth requisite'.[15] But for Hooke, there was never just one way to do something, and he also explained how his readers could make their own tiny microscope lenses out of a 'globule' of melted glass (which was effective but difficult to use); he himself, he said, had made other microscopes 'of Waters, Gums, Resins, Salts, Arsenick, Oyls, and with divers other mixtures of watery and oyly Liquors'.[16] He returned to this topic in the 1670s, spurred on by Antoni van Leeuwenhoek's discovery of bacteria and protozoa in 1676. In his 'Microscopium', printed as part of his *Lectures and Collections* in 1678, Hooke described again in detail how to make single-cell microscopes, adding that those who were not willing to take the trouble to make their own should go to Christopher Cock in Longacre as he had shown Cock how to make them.[17] Yet this was not the outcome that Hooke really wanted. He described his apparatus and the varied techniques he used to coax the best out of his instruments in such detail because he knew that to get good results an experimenter needed to understand his equipment intimately from raw materials to finished instrument. He wanted others to follow his example in not simply using such experimental apparatus but perhaps improving it as well.

Just as Hooke introduced others to Cock, Hooke's work with Cock also brought him into contact with other craftsmen and inventors. Some of these are nameless or unable to be identified, such as Cock's founder, who worked at a shop advertised under the sign of the Ball in Lothbury; or the 'ingenious glasse grinder' who was lodging 'at the boording Gentlewomens Scool in Hammersmith'.[18] Others were already known to Hooke. Through Cock he briefly resumed contact with Richard Reeve Jr. Reeve Jr had apparently taken over his father's business on the latter's death in 1666, but the Royal Society fellows considered him to be less skilful than his talented parent and patronized Cock instead. Reeve Jr stayed in the business, however, and in 1675 he was granted a patent for his 'art of casting and spreading of light by a new and unusual figure of foyled glass, pollished without grinding'.[19] Hooke visited his workshop on at least two occasions, and his diary suggests that he was moderately impressed by Reeve's method of joining glass plates for his reflector lamp. He speculated that 'the whole secret consists in the make and heating of the fornace and cooling it which is neer a week in doing. the grinding and bezelling the edges soe as to make a joynt being very easy.' On a later visit Reeve showed him the furnace, and rather unusually Hooke recorded a detailed sketch of the equipment, with explanatory notes (illus. 19). He ended on a less positive note, however: 'his glasses for spreading light were conicall foyled on the outside of the cone they spread not light soe well as a sphæricall concave'. As always, Hooke was interested in new techniques, but was immediately looking for ways to improve them.[20]

Hooke took an active role in directing work on instruments and components he commissioned from other craftsmen. In 1664 he wrote to Robert Boyle explaining that he had been slow to send him the ball and socket he had requested, 'for such have been the disappointments of those I had bespoke it of, that no

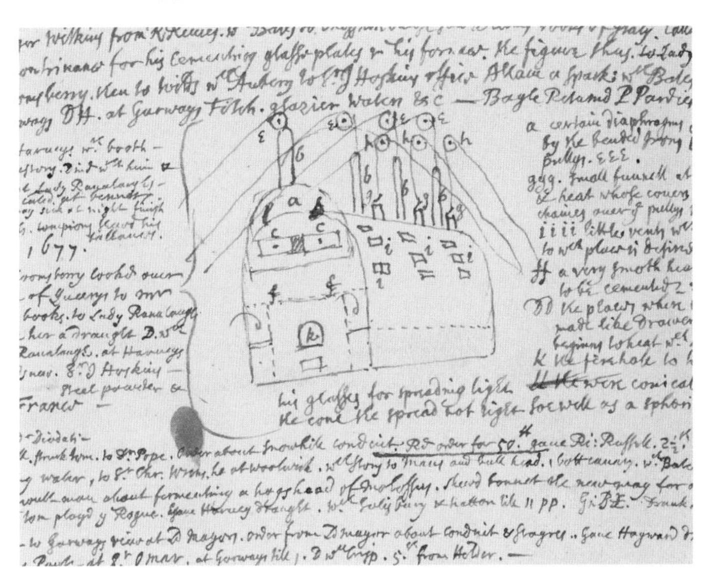

19 Richard Reeve's glassworking furnace, with explanatory notes, sketched in Hooke's diary (March 1677).

less than three have failed me, and I was fain to stand by a fellow most part of this day to direct him'.[21] Hooke's closest working relationship with an instrument-maker was probably his collaboration with Thomas Tompion. Tompion specialized in clocks and watches and came to be England's foremost clockmaker of the period. Hooke seems to have first come into contact with Tompion when he ordered a quadrant from him in April 1674. Although he initially featured in Hooke's diary as 'Tomkin', Tompion was clearly an immediate hit with Hooke. A few days after his first visit he returned to Tompion's workshop:

> to Thomkin in Water Lane much Discourse with him about watches. told him the way of making an engine for finishing wheels. and a way how to make a Dividing plate. about the forme of an arch. about an other way of Teeth work. about pocket watches and many other things.[22]

This avalanche of ideas suggests that Hooke realized he had met someone who was equally mechanically minded, and who might have the technical ability to help bring some of his inventions to fruition. Over the ensuing months the two enthusiasts had many more conversations about wheels, compasses, screws, springs, mills, steel, pumps, fire engines and lenses. Hooke was apparently at Tompion's when he 'Invented the way of printing with the common press pictures made with Pinns. An Invention of Great use.'[23] Sadly, no further details about this useful invention were recorded. Notes like these, however, suggest that Tompion's workshop was not simply somewhere Hooke went to order a specific instrument or part, it was a space in which he could be creative and inventive. Equally, many of these conversations took place in Hooke's rooms at Gresham College, where Tompion was a regular visitor on Sunday afternoons, often with Sir Jonas Moore, a mathematician with a keen interest in astronomical instruments. Hooke seems to have promoted Tompion's work among his circle of acquaintances, perhaps helping him to further commissions. He certainly praised Tompion in print, directing anyone who wanted a quadrant made to Hooke's new design to employ Tompion: 'this person I recommend . . . finding him to be very careful and curious to observe and follow Directions, and to compleat and perfect his Work, so as to make it accurate and fit for use.'[24]

Tompion continued to work with Hooke for the next twenty years. Their most intense period of collaboration began in February 1675, when Christiaan Huygens wrote to the Royal Society announcing that he had invented a new type of watch with a balance spring. Hooke immediately claimed that he had invented a watch with a spring mechanism a number of years previously (as indeed he certainly had done). Two days later Hooke was at Tompion's testing Huygens's spring, which he found to be 'not worth a fart[hing]'. This set off a flurry of activity, with Hooke instructing Tompion to produce various pieces of clockwork and visiting his shop frequently to check on progress. On 7 April Hooke, Moore and Tompion gained an audience with the king to show him Hooke's 'new spring watch'. According to Hooke, the king was 'most graciously pleased with it and commend[ed] it far beyond Zulichems [Huygens's]. he promised me a patent.' Hooke was not completely satisfied, however, and continued working with Tompion to test different springs, noting over the following weeks 'tryd perpendicular Sp[iral] Spring', 'tryd Double perpendigular spring' and 'the Thrusting Spring the best'. Occasionally he felt Tompion was not working quickly enough (writing 'tompion a Slug' in the privacy of his diary), but the two men clearly had a good working relationship, and Tompion trusted Hooke enough to ask his advice about a new, larger property he moved to the following year.[25]

Undoubtedly Hooke relied on instrument-makers such as Cock and Tompion to turn his ideas into concrete reality, but when we look at their interactions more broadly we can see that these relationships were equally significant for Hooke's philosophical method. For instance, one evening in October 1674 Hooke stopped in at Joe's coffee-house, where he 'discoursd with tompion about plug for wind pump. and about the fabrik of muscles. about Cork bladders Leather &c. about fire engine the way of making it.' Hooke had observed through his microscope the structure of muscles and leather and the tiny 'Boxes or Bladders of Air' in a slice of cork.[26] Even though he did not record how – or indeed whether – these microscopic structures in the natural world were connected with the wind pump and the fire engine, they were clearly part of his thinking about mechanical problems. This ability to slot ideas and observations from one field into an entirely different project was part of his brilliance. Modern historians have suggested that Hooke's knowledge of a range of mechanical devices enabled him to recombine parts in order to invent new mechanisms, pointing in particular to his equatorial quadrant with a clockwork drive and some of his suggested improvements to clock workings.[27]

Given his enduring devotion to instruments of all kinds, it is entirely fitting that the last publication to bear Hooke's name in his lifetime was a short description of his marine barometer, published by Edmond Halley in the *Philosophical Transactions* in 1701. Halley wrote the text because Hooke was apparently too unwell at the time to publicize his own invention: a double thermometer, or manometer, that allowed changes in the barometric pressure to be read.[28] Halley reported that the Hooke barometer he had taken on his own voyage to the South Atlantic in 1698–1700 'never failed to prognostick and give early notice of all the bad weather we had', adding 'from my own experience I conclude that a more useful contrivance hath not for this long time been offered for the benefit of Navigation.' More generally, he praised 'the great facility Dr Hook has always shewn, in applying Philosophical Experiments to their proper uses'.[29] For Hooke, theory and application were fundamentally intertwined. He argued that experiments and observations of nature must be linked with 'the Improvement of Art' (that is, human endeavours). Separately, he joked that these two strands of work were like warp and weft before they had been woven together: 'unfit either to Cloth, or adorn the Body of Philosophy'.[30] This was the philosophy he followed when working with instrument-makers, and as we will see, it also drew him into contact with some of London's leading artists and craftsmen.

Schem: XVIII
A
B
C
D
E
F
G
H
I

SIX

Curiosity and Beauty

When Hooke came to write about the tiny seeds of thyme he had been observing through his microscope, he felt it necessary to explain the accompanying illustration to his readers. The 'pretty fruits' they could see in the image, he wrote, were actually thyme seeds (illus. 20). They all looked slightly different, none quite the same shape, but they had one thing in common: 'they each of them exactly resembled a Lemmon or Orange dry'd.' Thus the seeds provided 'a very pretty Object for the *Microscope*, namely, a Dish of Lemmons' on a very tiny scale. Hooke was curious to see whether the insides of the seeds were constructed in the same way as the insides of lemons (which he thought would explain why their outsides looked the same), but he reported that they were merely a white pulp. Nevertheless, the relationship between the outside and inside was clear to Hooke.

> We may perceive even in these small Grains, as well as in greater, how curious and carefull Nature is in preserving the seminal principle of Vegetable bodies, in what delicate, strong and most convenient Cabinets she lays them . . . as if she would, from the ornaments wherewith she has deckt these Cabinets, hint to us, that in them she has laid up her Jewels and Master-pieces.[1]

20 Thyme seeds observed through a microscope, printed in *Micrographia* (1665).

Instead of being a prosaic reflection of interior structure, for Hooke, the beauty of the thyme seeds reflected the deeper significance of their 'seminal principle', their potential for growth. Perhaps what stands out the most in this description, however, is Hooke's eye for composition. Like the painters of still-life artworks in the seventeenth century, Hooke regarded a 'Dish of Lemmons' as an object that was at once both beautiful and instructive (illus. 21).[2] In *Micrographia* he saw beauty almost everywhere he looked. A piece of seaweed, he informed his readers, was incomparable among other plants for its beauty and 'curiosity' (intricate construction); he included its image to 'excite their curiosities' and to encourage them to collect and examine samples for themselves.[3]

When Hooke first arrived in London as a teenage boy, he contemplated a career as an artist. He had practised drawing as a boy on the Isle of Wight, and in London he apprenticed himself informally to Sir Peter Lely, the period's most successful painter of court portraits. However, he found that he could not tolerate the smell of the oil paints and so he left Lely and attended Westminster School instead. As his description and illustration of the thyme seeds suggests, he retained his painterly instincts. He was always alert to the significance of the visual, both for what it might say about materials or structures and for its potential to convey something more to the viewer: a deeper appreciation for the beauty and intricacy of the natural world. Hooke included drawing among the skills he felt were necessary for an experimental philosopher. The ideal philosopher should be able to 'design and draw very well, thereby to be able both to express his own Ideas the better to himself, to enable him to examine them and ratiocinate upon them himself, and also for the better informing and instructing of others'.[4] Drawing was not just about communicating information, it was a fundamental tool for thinking and learning,

for inventing instruments and devising experiments, as well as recording results.

Hooke's own aptitude for drawing and his time with Lely introduced him to the contemporary art world at a young age, but as he established himself as an experimental philosopher he met a much wider range of artists, craftsmen and followers of the many trades that flourished in and around London. In part, he was encouraged to do so by the Royal Society fellows. One strand of research that the society was keen to pursue in the early period was a 'history of trades'. Fellows were encouraged to observe tradesmen at work and send in written reports to the society. Like many of the society's early endeavours, this one was only partially successful, but some fellows did take up the challenge. The society listened with interest to papers on a range of topics including glass and cider production, tin mining in Cornwall, cheese-making in Cheshire and oyster-breeding in Colchester. Hooke contributed

21 Giovanna Garzoni, *Still-Life with Bowl of Citrons*, late 1640s, tempera on vellum.

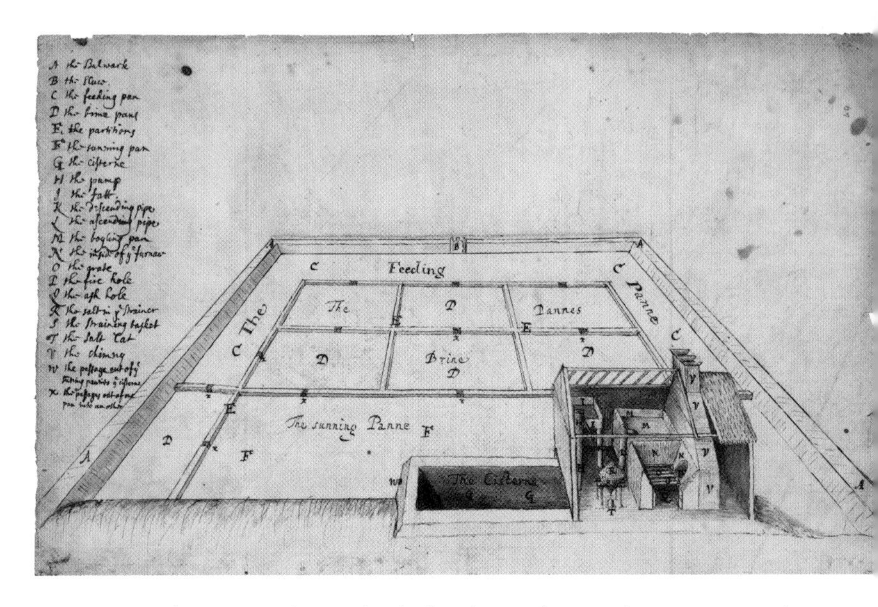

a lecture on the method of making salt at a saltern in Hampshire, perhaps something he had initially observed growing up on the Isle of Wight (illus. 22). Some fellows found this kind of research difficult: despite wanting to help, Hooke's colleague John Evelyn admitted privately to Robert Boyle that he disliked 'conversing with mechanical capricious persons'.[5] However, Hooke's own mechanical aptitude, and his intense curiosity about mechanical processes, may partly explain why he himself was perfectly happy to converse with tradesmen and observe them at work. In fact, his scientific method shows that he considered this to be a key area of research, on a par with investigation of the natural world. With remarks like Evelyn's presumably ringing in his ears, Hooke defended in detail what many of his non-scientific contemporaries would have seen as a rather pointless and certainly ungentlemanly interest in the daily activities of common tradesmen. He argued that useful information could be gathered 'even out of the most vile and seemingly most foolish and trivial things'. Unique and

22 Hooke's drawing of a salt-works in Hampshire, possibly on the Isle of Wight, 1666.

potentially instructive trade practices, he reminded his genteel audience, are performed by almost all 'Estates and Conditions of Men'. Real progress in scientific knowledge would require philosophers to 'take notice of, and enumerate all the Trades, Arts, Manufactures, and Operations, about which Men are imployed, especially such as either contain some Physical Operation, or some extraordinary Mechanical Contrivance . . . such as these will very much inrich a Philosophical Treasury'.[6]

To demonstrate what he meant, Hooke followed up on this comment by itemizing as many trades as he could think of, sorted into sections according to the branches of knowledge to which the trades might contribute. The result was a riot of early modern specialist activity. For example, Hooke suggested that the histories of the following trades might help to shed light on the subject of 'Earths and Clays':

> Potters, Tobacco Pipe-Makers, Glass makers, Glasiers, Glass-Grinders, Looking-Glass-Makers, or Foilers, Spectacle Makers, and Optick-Glass-Makers, Makers of Counterfeit Pearl and precious Stones, Bugle [bead] Makers, or Lamp-blowers, Colour Makers, Colour-Grinders, Glass-Painters, Enamellers, Varnishers, Colour-Sellers, Painters, Limners, Picture-Drawers, Makers of Baby [doll] Heads, and Bowling Stones or Marbles, Counterfeit Marble, Wax-work, Casters. Brick-makers, Tile-makers, Lime-burners, Plasterers, Paviers, Pargiters, Furnace-makers, China Potters, Crucible Makers.[7]

Even for someone so fond of lists as Hooke, this roll call might seem excessive, except for the fact that he himself would have been acquainted with tradesmen in many of these callings. He gathered other trades into groups linked by their association with metals, plants, animals, animal-products and human life, demonstrating that every employment, no matter how menial, had the potential to be instructive.

The focus on trade histories also pointed to an underlying principle that Hooke reiterated many times. To improve scientific knowledge, philosophers must not ignore the things that are 'most common, and therefore pass without regard, because usual'. Everyday matters can yield 'excellent' knowledge, when inspected closely for the first time:

> So that a diligent Naturalist can go no-where, but he may find a Subject for him to contemplate and examine, but especially in such Places as are most or least frequented, for the Obviousness in the one, and Difficulty in the other, has made Multitudes of considerable Observations to be neglected.[8]

There were no places more frequented than the narrow back-alleys and busy workshops of seventeenth-century London, but they were still relatively neglected by experimental philosophers.

But what would a philosopher do with these observations? For Hooke, they would be useful because, just as a knowledge of man-made mechanisms helped him to understand the mechanisms of the natural world, a knowledge of the productions of art would help to explain how materials and textures were produced naturally. There were two aspects in particular to consider: materials, their properties and how they were manipulated; and the instruments used in different trades – the 'various Mechanical Engines' that transformed raw materials into useful products. In his philosophical method Hooke listed all the techniques experimental philosophers could use to uncover the 'workings of nature'. He suggested 29 different operations in total. Of these, eighteen involved the close observation of nature, but the remaining eleven involved comparing nature with art, or human industry. These included suggestions such as observing the ways in which art can cause nature to change, for example, gardeners using artificial heat to ripen fruit; or observing natural and artificial ways of producing the same effect, such as the incubation of eggs either in ovens or naturally. Knowing that it was possible to hatch eggs by providing the right level of heat in an oven meant that the mind was not tempted to imagine any 'strange kind of irradiating and plastick Influences from the brooding Hen to the hatching Eggs'. He suggested that 'Paper or Hats' could be usefully compared with animal or human skin, because his observations of the texture of paper and the felted wool used to make hats seemed similar to the texture of skin. As he explained, paper, felt and skin were not exactly the same in every way,

> yet there are many [details] that seem to have a great Affinity, and serve to prompt the Intellect, and very much to help the Fancy and Imagination to conceive of those things, and of the Method of Nature, and they will serve greatly to instruct the Mind what things are to be look'd after and examin'd in the Proceedings of Nature.[9]

He went on to argue that the more examples of relevant trade processes that could be compared, the better. The philosopher who saw that different ways of working with a particular material could result in a similar output would be prompted to ask more questions and would be less likely to jump to the lazy conclusion that because the natural process and the artificial process were similar in one way, they were similar in all ways. Therefore, Hooke wrote, 'for this purpose it would be very requisite to have a perfect Account of all the Productions of Art, such as are dispers'd up and down in several Trades and Occupations of Men, whether for Profit or Pleasure.'[10]

Hooke probably chose the example of comparing felt hats with skin because the hat-making trade was another that he had observed himself and reported on to the Royal Society in February 1666 (illus. 23). His report detailed all the steps that the felt-makers, whom Hooke called 'artists', went through to make sheep's wool into felt hats. He described their special tools and other equipment, and listed the materials and substances that they used in the different stages of the lengthy process. The description shows that Hooke took his own advice about the importance of paying attention to common things, and indeed things that might be considered 'vile': one of the stages in felt-making required the use of urine to treat the wool. Hooke even described the special shirt-cuff worn by felt-makers so that their sleeves did not get caught in their tools. No detail was too small to record. Perhaps the most significant aspect, however, is the amount of time that Hooke must have spent with the felt-makers, discussing their work and watching them in action.

23 Hooke's drawing of felt-makers at work, *c.* 1666. The instructions written on the drawing show that it was intended to be printed, but it does not seem to have been.

The impression we gain is of a man who was willing to take manual labour, and those who work at it, seriously, and one who was able to win the confidence and respect of these artists as they shared their trade secrets with him.

This was not an isolated instance, and there is evidence scattered throughout Hooke's diary of his interactions with London's artists and craftsmen. Sometimes he went to watch and learn, as in his visits to the foundry with Christopher Cock, for example, or over the years to various glass-houses, potteries, a lead-mill, 'the Scarlet Dyehouse' and, a few days before Christmas in 1678, to a 'ribbin and silk stocken weaver' with John Houghton and Francis Lodwick. Often there was some specific point to the encounter, relating either to his work for the city or the Royal Society, or to other business. For example, over the years he engaged various engravers to prepare the illustrations for his own books, but also occasionally to work on other books, such as Moses Pitt's projected eleven-volume atlas (of which only four volumes were eventually published), or on completely different projects, such as commissioning a suitable painting and engraving of Bethlem Hospital after its completion in 1677. In terms of his own printed books, he took an active role in creating the engravings, correcting plates as they were being completed and overseeing technical details such as the addition of lettering to provide an explanatory key to the images. Collaborating with the engraver Robert White on the plates for *A Description of Helioscopes* in 1675, Hooke noted in his diary that he had 'shadowed [the] plate', that is, sketched out an image; later he met White in Bloomsbury and gave him the 'amendment' of the plates. It was White who produced an engraving of Bethlem Hospital, rather than David Loggan, who had been Hooke's initial suggestion. For his next book, *Lampas*, Hooke worked with a different engraver, William Sherwin, delivering to him the 'schemes' he had drawn for a plate and later returning the 'rectified' plate itself. Sometimes multiple people were involved. On one occasion Hooke employed his assistant, Harry Hunt, to engrave his plates. Hunt was an accomplished draughtsman and was normally on extremely good terms with Hooke, but he seems not to have relished this commission: Hooke noted 'Harry finish plate but grumbling. put plates to Lamb to Letter.' Francis Lamb later engraved plates for further projects in which Hooke was involved, including Pitt's abortive English atlas.[11]

The primary reason for interactions like these was to commission work, but Hooke and his Royal Society colleagues were always fascinated by new materials and techniques, in art as in other areas. Prince Rupert, Charles II's cousin and a Royal Society fellow, is credited with either inventing mezzotint engraving (as John Evelyn thought) or introducing the new process to England in the 1660s. Hooke recorded in his diary several conversations with friends about printing, and some of his own ideas and suggestions. A method of printing pictures with pins was one invention, but he also noted an idea for 'etching upon horne with a needle' and new ways of engraving lettering.[12] In March 1679 he told Sir John Hoskins about his 'contrivance for tinplates for Rolling presse. and my method for printing books.' He had similar conversations in the same week with Wren and Moses Pitt, recording that Pitt had agreed to his 'contrivance for printing books'. This was the period when Hooke, Wren and other Royal Society fellows were collaborating on Pitt's ambitious atlas, and Hooke's suggestion was intended to help with the printing of this vast work. Wren was equally interested in printing techniques, and Hooke's diary mentions in passing Wren's method of mezzotint using 'powder', and of printing with a 'hobby plate' (a thin brass plate). A conversation at Jonathan's coffee-house yielded a note: 'German paper that will print dry.'[13]

As always, the emphasis was on usefulness. This kind of printing normally required damp paper to help the ink transfer, so paper that would print dry would speed up the process. Hooke acknowledged that there were two potential audiences for descriptions of trades. One was a philosophical audience, who would use the information for 'Philosophical Inquiry, for the Invention of Causes, and for the finding out the ways and means Nature uses, and the Laws by which she is restrain'd in producing divers Effects'. The other was an audience of people who knew little about the trade or process, but who wanted to know more for their own purposes, 'for Curiosity, or Discourse, or Profit, and Gain, or the like'.[14] In his own description of felt-making, read to the Royal Society, he set out several aims:

> In delivering you a history of this Art I shall first explaine their tooles and Instruments. Next their materialls and manner of working and thirdly I shall indeavour to draw some inferences from my observations and shew what information they afford us for the finding out the operations of Nature. Lastly some conjectures or attempts how this art may be varyed or improvd either as to the materialls on which they work or as to the instruments and manner of their working or both.[15]

Although he did not go on to draw any inferences here about the 'operations of Nature', or suggest ways in which the felt-makers' processes could be improved, both of these aims were given equal weighting in his report. Hooke's words here hint at an unstated but significant reason for investigating trades, aside from scientific knowledge: financial gain, either for individuals or more broadly as an improvement to England's manufacturing and trade.

Hooke was not alone in his drive to innovate, and it was not only tradesmen who wanted to improve manufactures. Among

Royal Society fellows, Charles Howard, brother of the Duke of Norfolk, applied for patents for new methods of tanning leather and processing flax; the well-connected Herefordshire clergyman and landowner John Beale published a number of treatises on orchards and cider-making; others investigated glassmaking, dyeing and ceramics. At Jonathan's coffee-house in 1692 Sir John Hoskins showed Hooke and his companions a sample of 'flock work printed in flowers' made with powdered mica from Muscovy passed under a hot printing roller. Hoskins was a lawyer and a baronet, and had served as MP for Hereford, but he clearly enjoyed experimenting: two weeks later he told Hooke that he had improved his printed flock work. Hooke himself had experimented with a process of printing onto cloth almost twenty years earlier. He had done this in the company of Patrick Barrett, a manufacturer of printed cloth based in Moorfields. Barrett must have been producing high-quality wares, as on one occasion Hooke noted that he had seen his 'quadruple printing', which he judged to be 'very fine'. In March 1674 Hooke and Barrett tried 'Golding [a] flowered Shift' – that is, presumably, adding a pattern in gold to cloth already printed with a flower motif. Hooke visited Barrett's workshop many times, often noting materials or new products, such as his 'new painted hanging' or his 'copper scouring earth'. On at least one occasion he seems to have commissioned Barrett to cast something in metal from a pattern Hooke supplied. Barrett was a member of the Worshipful Company of Blacksmiths and was thus in a good position either to do this work himself or arrange for someone else to do it.[16]

Hooke and his companions did not see these activities as separate from their philosophical enquiries. Hooke sometimes met Barrett in company with his Royal Society colleagues and other times took his friends to Barrett's workshop. Having seen Barrett's abrasive 'earth' for scouring his copper plates, Hooke asked at the Royal Society meeting the following day whether anyone 'could give any information concerning a certain English earth very effectual for scouring copper, brass, etc'. None of the fellows had any further information, but at the following week's meeting Hooke himself brought in a sample of it, which he said had come from 'Lancashire, Derbyshire or Cheshire, or that way'. Clearly his informant, possibly Barrett, had only been knowledgeable up to a point.[17]

Hooke and his Royal Society colleagues were always interested in local natural products, particularly those that might be valuable or useful. Hooke shared this interest with the potter John Dwight, whose sculpted ceramic figures impressed Hooke in the 1670s. The two men had met at Oxford, where Dwight studied civil law and may have worked in Robert Boyle's laboratory. Dwight had swapped his career in law to become an experimental chemist and had taken out a patent to make 'transparent Earthenware commonly known by the names of Porcelane or China and Persian Ware as also . . . the stone ware vulgarly called Cologne ware'.[18] He was attempting to copy valuable imported goods, and Hooke and his colleagues were naturally curious about his process. In September 1673 Hooke noted that Dwight had told him 'he used salt to throw into his fire' – a reference to the method of glazing by throwing salt into a kiln. Hooke was impressed. A few months later he wrote: 'Saw Mr Dwights english china. Dr Willis his head, a little boye with a hauke [hawk] on his fist, severall little Jarrs of severall colours all exceeding hard as a flint, very light, of very good shape. the performance very admirable and outdoing any European potters.'[19] In February 1675 Hooke showed an example of Dwight's work, a bust made of Dwight's 'Porcelane' (not, at this point, a true porcelain), to representatives of the city and the Royal College of Physicians. The previous week the physicians had asked Hooke to commission a bust of their benefactor, Dr Baldwin Hamey the Younger, from the sculptor Edward Pearce. Clearly Hooke thought that Dwight's porcelain might be an alternative to the more usual marble. Once again, Hooke passed on information about these processes to the Fellows of the Royal Society. He brought the portrait bust into a meeting to show his colleagues. The minutes note that it was 'made in England, of English clay, so hard and solid, that [Hooke] said, nothing would fasten on it, except a diamond'.[20] Its solidity, and perhaps its resistance to scratching (except by diamonds), must have recommended Dwight's porcelain to Hooke for a portrait bust to be displayed in a busy institutional setting. Further references in his diary suggest that Hooke may also have recommended Dwight to officials in the city as someone who could provide water pipes for London's upgraded conduits.

Hooke did not himself go on to patent any inventions, nor did he make a fortune from improvements to any of the trade practices he so assiduously observed. To understand the significance of his involvement in trades we need to look more closely at his work. Like his aptitude for mechanics, which allowed him to recombine individual parts into new devices, his understanding of trade processes and materials prompted new ideas. In *Micrographia*, in a section discussing a piece of taffeta silk ribbon, he described having seen a 'pretty kind of artificial Stuff' that was transparent but glutinous when wet, and thus easily dyed: 'to the naked eye, it look'd very like the substance of the Silk,'

> And I have often thought, that probably there might be a way found out, to make an artificial glutinous composition, much resembling, if not full as good, nay better, then that Excrement, or whatever other substance it be, out of which the Silk-worm wire-draws his clew [silk thread]. If such a composition were found, it were certainly an easie matter to find very quick ways of drawing

> it out into small wires for use. I need not mention the use of such an Invention, nor the benefit that is likely to accrue to the finder, they being sufficiently obvious. This hint therefore, may, I hope, give some Ingenious inquisitive Person an occasion of making some trials, which if successfull, I have my aim, and I suppose he will have no occasion to be displeas'd.[21]

This suggestion drew on Hooke's understanding of the properties of silk, his experience of dyeing different materials and his knowledge of the process of making very fine strands of wire. It is entirely typical that Hooke described the silkworm's spinning of silk filaments in terms of the mechanical process of wire drawing. None of Hooke's 'Ingenious inquisitive' contemporaries acted on his hint, however, and it was not until the late nineteenth century that artificial silk came onto the market.

Only a very few artificial objects are described in *Micrographia*, since according to Hooke, 'the Productions of art are such rude mis-shapen things, that when view'd with a Microscope, there is little else observable, but their deformity.' There was nothing particularly instructive in the microscopic detail of most man-made productions, unlike the products of the natural world, which revealed more beauty and intricacy the closer one could look. The investigation of taffeta was an exception to this rule; watered silk was another. It intrigued Hooke sufficiently that he took the time to explain its peculiar properties. The reason that the silk changed colour depending on how it was viewed was because the threads were 'by the Mechanical process of watering, creas'd or angled in another kind of posture then they were by the weaving', and this caused light to reflect from different parts of their surface. To understand this further, Hooke suggested that 'here we must fetch our information from the Mechanism or manner of proceeding in this operation; which, as I have been inform'd, is no other then this.' He went on to describe the watering process in detail so that his readers could understand for themselves how it affected the finished cloth. Being Hooke, he added that the reflections he had observed in the watered silk could also explain why 'a small breez or gale of wind ruffling the surface of a smooth water, makes it appear black', or indeed 'multitudes of other phaenomena' where light is reflected from an irregular surface.[22]

Passages such as these show that Hooke's enthusiasm for understanding crafts and trades was a productive part of his scientific method, but one suspects that in the eyes of many of his colleagues it was the possibility of making money from an improved or new process that was most alluring, rather than pure research. In 1692 John Houghton FRS, a London apothecary nominated by Hooke for fellowship of the Royal Society, began publishing a weekly periodical, *A Collection for Improvement of Husbandry and Trade*. Houghton's paper is best known for being the first to print the current prices of agricultural products, manufactures, imports and exports, and stocks in London companies – essentially the first stock exchange figures – but he also promised to print 'Ample and exact Histories of Trades, as Maulting, Brewing, Baking, Tanning, Dying, Potting, Glass-making, and many others'.[23] He was following in the footsteps of another fellow of the Royal Society, the printer and globe-maker Joseph Moxon, who had published detailed illustrated descriptions of the trades of printer, smith, joiner, carpenter and turner. Moxon and Houghton were both unusual among the Royal Society fellowship because they were tradesmen: most fellows were gentlemen or noblemen. As we have seen, other fellows participated in specific trade practices, but with the exception of Christopher Wren, they were not members who contributed to what we would traditionally consider as scientific advances.

Hooke was unique among his scientific peers in being willing to learn from skilled tradesmen at first hand, and at the same time being able to put his knowledge to use in the service of his scientific work. Remarkably, his philosophical method gave equal weight to knowledge of the natural world and knowledge of 'Mechanical Employments and Operations'. In fact, he wrote, even though in his method he had dealt with investigations into natural phenomena first and artificial phenomena second, it was the latter that should 'primarily and chiefly' be studied, because it was knowledge of these that would guide the mind to ask the right questions and make fruitful investigations. He argued persuasively and at length in his philosophical method that tradesmen's workshops were a kind of laboratory and that useful knowledge could be found in them. His collaborations with artists and craftsmen remind us again of the physical, practical nature of Hooke's approach to science. Through Hooke's activities, we can get a glimpse of a diverse cast of characters who, although not experimental philosophers themselves, were nevertheless crucial supporters in the development of science in Restoration London.

SEVEN

An Excellent System of Nature

ith all this activity buzzing around him, it is perhaps not surprising that Hooke took such pains to note the details of every day's meetings and conversations in his diary. His philosophical method frequently mentioned memory: along with the bodily senses and the faculty of reason, it was one of the three human attributes that enabled new knowledge to be produced. Unfortunately, Hooke noted, 'the Memory is frail, and may quickly forget even those things that are of most Importance, and does not without much Labour and Trouble at best, recal all Circumstances that are considerable at the time when they are most requisite.'[1]

Hooke mistrusted his own memory, and even his close friend John Aubrey remarked that although Hooke's 'Inventive faculty' was prodigious, his memory was not: 'for they are like two Bucketts, as one goes up, the other goes downe'.[2] Hooke may have tried to remedy this with medical preparations. In 1677 he noted that an acquaintance had told him about a friend of his who had recovered from a bad memory ('and severall other Distempers') by dosing himself occasionally with a small amount of very fine silver filings.

Even though Hooke's work as city surveyor was slowly diminishing during the later 1670s and early 1680s, he was still busy with private architectural commissions, his Gresham College duties and his roles at the Royal Society. His daily journal entries, which had been constant throughout the previous decade, started to tail off in 1681 and ceased completely in May 1683. Long runs of date-only entries suggest that it was not a deliberate decision to stop writing his diary, so perhaps the press of business had finally distracted him from his daily records. His other papers show that, as always, he was working on a range of subjects at the same time. Among these are a series of lectures on the subject of light, delivered at Royal Society meetings in the early 1680s. This was not a new topic for Hooke. He had already announced his theory that light travelled in waves or pulses in *Micrographia*, using it to explain the production of colours in thin transparent bodies, such as pieces of mica. In the 1670s he had argued with Newton about the nature of light and colour, rejecting Newton's particle theory of light (Hooke believed it was created by 'pulses', or waves, travelling through the aether) and his theory, obtained via the famous prism experiment, that white light was a compound of different coloured rays. In this later series of lectures Hooke restated some of his previous ideas, argued that light is 'purely corporeal', and thus subject to the same physical laws as all other tangible bodies, and stated that the propagation of light acts according to an inverse square law: that is, its intensity is inversely proportional to the square of the distance from the light source. At the end of these lectures, however, he deviated into what was unusual territory for him. Prompted by his musings about the motion of light and the human perception of time, he offered his own theory of how the human brain, and in particular memory, functioned. Historians have largely treated this lecture as an oddity among Hooke's work (which it is). The brain had of course been discussed previously by early modern philosophers – in particular Hooke's early Oxford employer, Thomas Willis, who had published a ground-breaking book about the anatomy of the brain with detailed illustrations by Christopher Wren (illus. 24). Hooke, however, put forward a mechanical explanation for

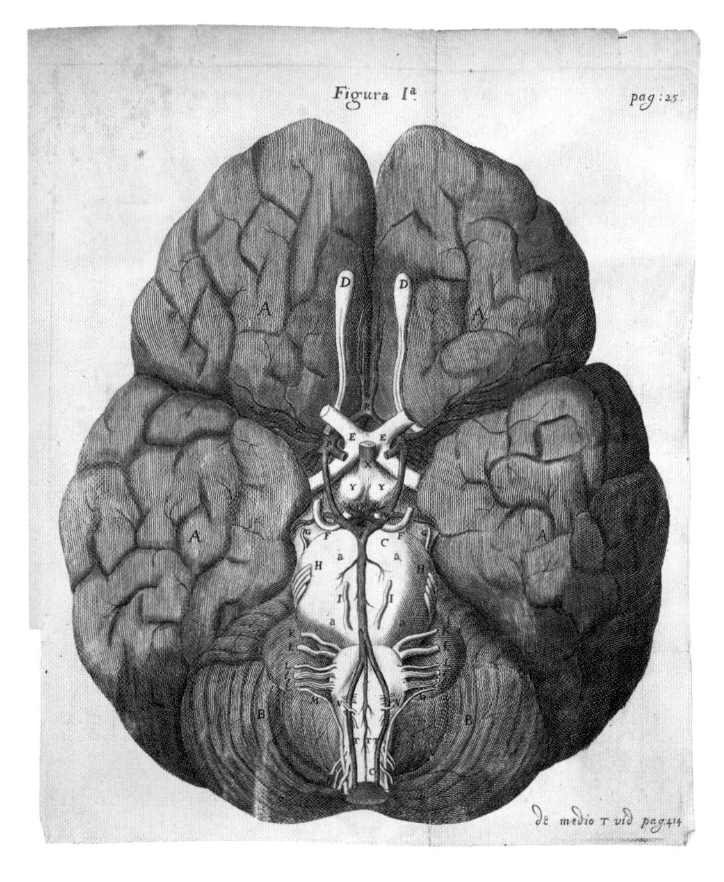

human thought processes that was unlike anything his contemporaries had suggested. Ultimately, his theory of memory explains why Hooke thought his scientific method worked.

Based on his microscopy, Hooke argued that humans are only aware of time on a human scale: surely time must feel different for a tiny fly, beating its wings in a pattern too fast for humans to comprehend. He suggested that human memory was an organ for sensing time, just as the eye was an organ for seeing. But how did memory work? Like other philosophers, Hooke believed that

24 The human brain, from Thomas Willis, *Cerebri anatome* (1664).

information about the physical world entered the body via the senses. When it was transmitted to the brain, it formed memories. Hooke suggested that each memory was in fact a physical thing: a tiny particle formed from some unknown substance in the brain that was capable of storing sensory impressions. The brain constantly created new memory particles, or ideas, as the body experienced new sensations. These particles spiralled out from the point where they were created, in a 'Chain of Ideas' coiled up in the brain.[3] The distance between the central point of creation and an idea's position in the chain indicated the amount of time that had passed between that idea's creation and the present moment.

Hooke was satisfied that this basic structure explained the human capacity for gauging the passage of time, but his model suggested some other possibilities to him. Since the idea particles were real, and solid (though very tiny), he assumed that they acted like any other tiny particles of matter: they had a shape, and they were always in motion. They also interacted with one another. When a new sense-impression formed an idea that shared some similarities with older ideas further away in the chain, Hooke suggested that these similar particles would resonate together, and that this would prompt the soul to form a new idea. New ideas could interact with older ideas of a similar nature because they vibrated in harmony with them, and this could renew the older memories: their vibrations would attract the soul's attention, recalling them to the mind. But particles could also jostle together until they lost their original shape and became worn, or were pushed to the outer edge of the coil or chain of ideas and lost, and this explained why some things were forgotten.

Overlooking all this activity was the soul. When Hooke described the soul's interaction with the 'Organ of Memory', he used the analogy of the Sun in the solar system. Just as the Sun's rays spread out and touched everything, so the soul's awareness connected it with each idea. It was the soul that was conscious of the passage of time, and the soul that performed the action of thinking, by focusing 'a more particular Radiation' on to specific ideas in the chain of memories and at the same time framing new ideas.[4] Intriguingly, Hooke speculated that the soul might even be aware of ideas outside its own body, and thus able to read other people's thoughts. But in bringing the soul into his theory Hooke encountered stiff opposition from his colleagues. From the early days of the Royal Society the fellows had agreed that they could discuss any topic except 'God and the Soul', which remained strictly off-limits to experimental investigation. Speculating about its role in human thought processes brought the soul uncomfortably into the realm of the concrete: the minutes of the Royal Society meeting record the objection that Hooke's theory 'seemed to tend to prove the soul mechanical'.[5] The problem with a material soul was that it seemed to call into question the existence of the spiritual realm, and therefore the existence of God. Hooke's primary allegiance was to philosophy rather than theology, so this may not have bothered him too much, but he apparently dropped the subject rather than risk antagonizing his contemporaries further. His lecture was only published after his death.

The fact that he never returned to his lecture on memory and cognition suggests that Hooke may not have taken all these ideas too seriously. And yet it seems unlikely that he would have come up with a theory that did not fit his own experience of his thought processes. What might it reveal to us about Hooke's unique mind? Two things that stand out are the explicitly material nature of the memory particles and the (mostly) mechanical process Hooke described of remembering and thinking. Hooke envisaged the human brain as a piece of clockwork in which particles were bound together by the same laws of attraction that govern celestial bodies. Even the soul operated according to the laws of physics: it acted more powerfully on ideas that were nearer the central point of creation than those further away, 'in a duplicate proportion to their Distance reciprocal' – that is, according to the same inverse square law Hooke had proposed for light and gravity.[6] In this clockwork conception of human thought, if everything was working correctly, the brain should automatically produce new ideas from the stimulations experienced by the body, just as Hooke believed that his philosophical algebra would lead inexorably to the correct outcome if the original data was sound. The third aspect that stands out is Hooke's concept of ideas as physical particles stamped with sensory impressions. The more a body experienced (tasting, smelling, seeing and so on), the more ideas were produced, and the more likely it was that they would interact with other particles to produce new ideas. Hooke even included a long digression in which he tried to calculate exactly how many idea particles might be stored in the brain, suggesting a total of 1,826,200 for a person aged fifty (with the proviso that some men might produce more, some less, according to how active their soul was). Thus Hooke's model of the human mind provides a direct rationale for his insistence on the value of watching tradesmen at work, of experimenting with materials and of observing even common and everyday substances, that is, exactly the kinds of research activities we have seen him doing. It also explains why he was so optimistic about the power of experimental philosophy to uncover new knowledge: with this kind of mechanical brain at work, it was hardly possible not to think new thoughts.[7]

As we have seen, despite this optimistic vision of the brain's workings, on day-to-day matters Hooke distrusted his own memory. Aside from any possible medical interventions, it is not surprising that he took steps to ensure that his forgetfulness did not hinder his work – or the work of other philosophers.

In his philosophical method he described an ideal, but somewhat idiosyncratic, procedure for scientific record-keeping that in many ways mirrored the processes that he suggested were taking place inside the brain. He insisted that all the details of an experiment or observation should be recorded as soon as possible, even aspects that may not seem relevant at the time, because they may turn out to be important when investigating a different topic. The details should be noted in the briefest format possible (Hooke suggested shorthand) on small individual pieces of 'very fine paper'. These pieces of paper should then be stuck onto a single page in a large book, so that all the details of an experiment could be seen in one view. Hooke suggested that the papers should be stuck down with 'Mouth Glew', a compound made of sugar and isinglass (a substance obtained from the swim bladders of fish) that became sticky when moistened with the tongue. This weak glue meant the papers could be taken up again and reordered, like modern Post-it notes, or removed completely if new information needed to be added. Just like the memory particles inside his brain, Hooke envisaged these snippets of information jostling and replacing each other or coming together to form a new insight when viewed in a single glance. Recording data in this way made it 'material and sensible . . . impossible to be lost, forgot, or omitted', and helped the brain in the same way that scientific instruments helped the senses. We do not have any examples of notebooks compiled using this method, but they would have been fragile and ephemeral, so perhaps this should not be surprising.[8]

His paper model of memory was presumably formulated in the late 1660s when he wrote his philosophical method; his lecture on memory was delivered in June 1682. The two models are remarkably consistent. Interestingly, they are also consistent with another theory that Hooke developed throughout his life, one which became a fundamental part of his physics. This was

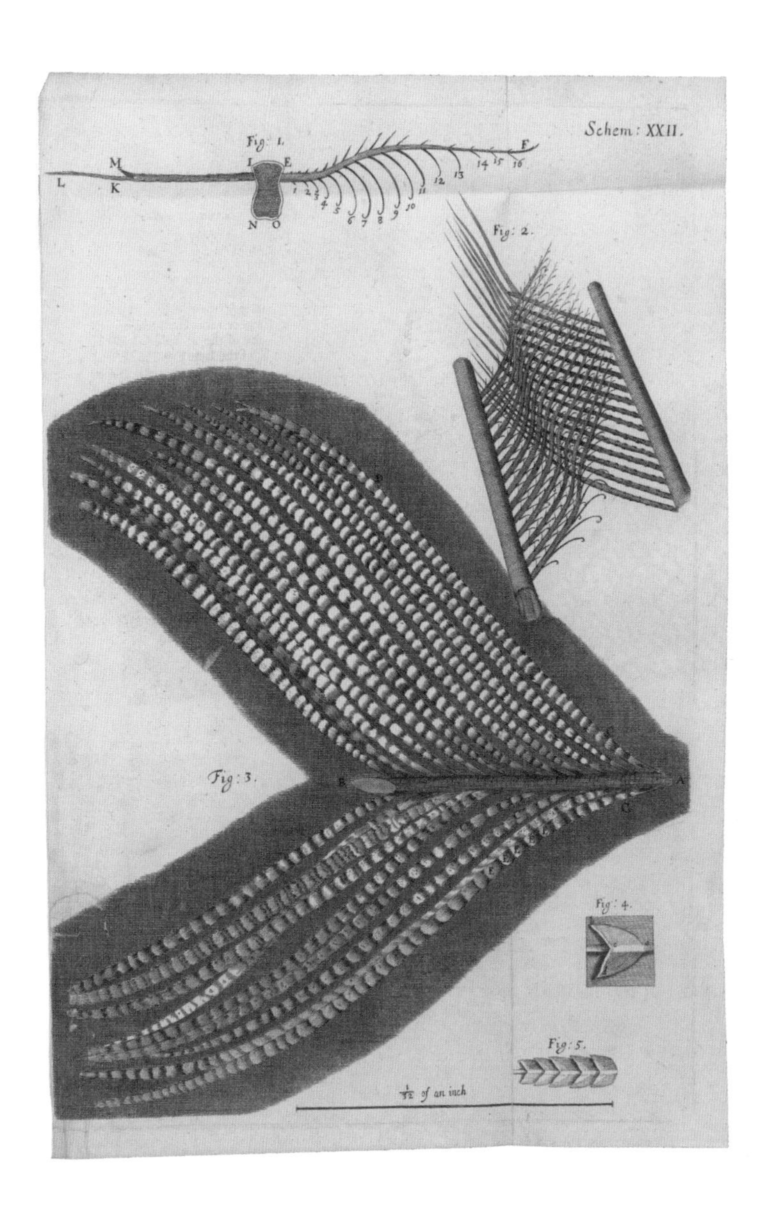

25 Microscopic details of peacock feathers, illustrated in *Micrographia* (1665). Hooke suggested that the colours changed when the feathers were wet because the water had more congruity with the structure of the feathers (fig. 3) than the surrounding air.

the theory of congruity, or the idea that some (congruous) fluids would easily mix with or stick to other fluids or solid bodies and some (incongruous) fluids would not. In Hooke's model of the brain ideas that were similar vibrated in harmony, and when the soul perceived several ideas acting in harmony it produced an 'Idea of greater Perfection', with aspects of all the previous harmonious ideas. Particles vibrating in harmony were thus congruous. Hooke had introduced the idea of congruity back in his 1661 work on capillary action to explain the fact that when narrow glass tubes with open ends are partially submerged in water, water rises in the tube above the level of the surrounding water. He suggested that the water was more congruous with the glass than air was, which made it harder for the air to get into the tubes. He expanded his theory further in *Micrographia*, where he returned to the topic of narrow glass tubes but also mentioned congruity and incongruity in connection with a range of seemingly disparate phenomena: the toughness of animal, vegetable and mineral substances; the globular form of sparks struck from steel; the appearance of colours in saline liquids; the pain arising from nettle stings; methods of staining marble, wood and cloth; and the colours in peacock feathers (illus. 25).

Congruity was relevant to all these topics because it was universal. Hooke suggested that it was linked with fluidity, which he argued was 'nothing else but a certain pulse or shake of heat' that loosened the particles in a substance so much that they moved and became fluid. He explained this by referring to a dish of sand being constantly shaken, which made the sand act as though it was liquid. Imagining that the small particles of matter in a fluid acted a bit like the grains of sand, Hooke pointed out that if the dish contained different sands of varying sizes, the larger grains would be ejected by the fine sand and would gather together. He suggested that a process like this took place in fluids:

> particles that are all similar, will . . . vibrate together in a kind of Harmony or unison; whereas others that are dissimilar . . . [will] make quite differing kinds of vibrations and repercussions, so that though they may be both mov'd, yet are their vibrations so different, and so untun'd, as 'twere to each other, that they cross and jar against each other, and consequently, cannot agree together, but fly back from each other to their similar particles.[9]

Particles that moved in the same way were congruous, and particles that were dissimilar vibrated differently and were incongruous. Hooke expanded further on the theory, explaining how different properties of particles (size, shape and material) affected congruity and incongruity.[10] He also used other 'similitudes' to illustrate his theory, most frequently the idea of musical strings vibrating and creating harmony or discord. He argued that solids interact with fluids because solid matter also vibrates; in fact, everything in the universe must vibrate, he reasoned, because everything has some degree of heat. Congruity was the cause of some materials' toughness, and it also explained magnetism.

Over the next decade Hooke continued to refine his theory of congruity. He mentioned it in passing at Royal Society meetings while discussing a range of subjects, including his experiments on the formation of bubbles from soap and water, the appearance of fog or mist in the air and the solubility of different salts. It had also captured the interest of Isaac Newton, who was working in Cambridge but periodically sending research papers to the Royal Society. After reading *Micrographia*, Newton began referring to matter as 'sociable' or 'unsociable', terms that closely resembled Hooke's.[11] By the time Hooke published his lecture on springiness in 1678, congruity had for him come to form a fully integrated theory of nature. After setting out the law of springs, Hooke explained that the property of springiness was caused by congruity. He gave a revised definition, now stating that it was 'nothing else but an agreement or disagreement of Bodys as to their Magnitudes and Motions'. There was no further mention of particles' shape or material. Instead, congruous bodies had 'the same Magnitude, and the same degree of Velocity, or else an harmonical proportion of Magnitude, and harmonical degree of Velocity'; incongruous bodies had neither the same magnitude or velocity nor a 'harmonical proportion' of magnitude or velocity. Bodies were held together by congruity because their particles were moving in harmony with each other, but differently from the bodies or fluid around them, and the incongruous movement of the 'ambient bodies or fluid' kept the congruous body together.[12]

In his lecture on springs, Hooke used the theory of congruity to explain the property of springiness, but in order to do so he had to posit the existence of a 'fluid subtil matter', or aether, that surrounded all solid bodies and held them together by vibrating at a different velocity from the particles in the solids. When springy bodies were compressed, their vibrating particles collided more frequently, which caused the body to try to spring outwards back into its original shape; when bodies were stretched, there were fewer internal collisions, and they sprang back inwards. In 1661 he had suggested that this subtle encompassing fluid, the aether, might also cause gravity by pushing all 'earthly bodies' away from itself and towards the centre of the earth; however, he somewhat spoiled the effect of this by immediately claiming to know 'a more likely hypothesis for gravity', which he could show by experiment. In conversation with many of his contemporaries, Hooke continued to work on questions about gravity and related problems such as magnetism, static electricity and the propagation of light. In a lecture delivered to the Royal Society in 1683 he tackled gravity again, attributing it to a 'vibrative motion' in the earth that was communicated to the aether, pulsing outwards in an orb. The aether permeated all solid and fluid bodies, and the motion it transmitted from the earth forced the bodies within its sphere of influence downwards towards the earth's centre. He acknowledged that it may 'perhaps seem a little strange how the Propagation of a Motion outward should be the cause of the Motion of heavy Bodies downwards'. To help his audience understand how this might work, he mentioned 'an Observation very commonly known amongst Tradesmen': the method of fixing an axe or hammer head onto a handle by holding the tool vertically with the head hanging downwards and the handle at the top, and hitting the top end of the handle, making the head work its way up the handle shaft.[13] In this analogy, the axe head was the body being drawn towards the earth's centre, and the handle was the fluid aether transmitting the vibrations from the earth.

As always, a known procedure in the world of the senses (fitting the head onto an axe) could be used to explain something that was outside the sensory realm. Hooke's lecture is probably incomplete, but he clearly expected 'Objections' against his hypothesis, one of which was that the earth's internal vibrations were impossible to detect. In a way, however, he had already answered this question. In an earlier lecture about the appearance, constitution and causes of comets, and the existence of aether, Hooke hinted at the limits of his method:

> this possibly may be the utmost that Man's Senses and Reason will ever inable him to perform, in the acquiring of the Knowledge of such Causes, Principles and Operations; the Method and Instruments wherewith they [comets and aether] work being far removed beyond the reach of our Senses: And therefore the best

> and utmost we can do towards the discovery of them, is only accurately to observe and examine all those Effects produced by them, which fall within the Power of our Senses, and comparing them with like Effects, produced by Causes that fall within the reach of our Senses, to examine, and so from Sensibles to argue the Similitude of the nature of Causes that are wholly insensible. And this is the utmost Bound and Limit of our most exalted and regulated Reasoning, beyond which that Power cannot carry us.[14]

He went on to reassure his audience that if a 'right Method' was used, the power of human reasoning would still reveal more than had ever been imagined; but unlike the optimistic young man who believed it might be possible to hear sounds from other planets, Hooke now acknowledged a boundary to the experimental philosophy.

Hooke's colleague Isaac Newton, meanwhile, was preparing a book that in many ways fulfilled Hooke's prediction of a new 'System of the World' that he had challenged astronomers to find in 1674.[15] The manuscript of Newton's *Principia* was initially presented to the Royal Society in April 1686. In it, Newton determined the inverse square law of gravity – an idea that Hooke believed he had communicated to Newton some time earlier. He immediately pointed this out, causing Newton to respond angrily (to Edmond Halley, who was acting as intermediary) and remove a reference to Hooke from the printed version of the book. As might be expected, there was some truth on both sides. Scholars agree that Newton changed his mind about gravity after exchanging letters with Hooke in 1679. During this correspondence, which Hooke initiated, Hooke asked for Newton's opinion on planetary motion and made the point that gravity operates according to the inverse square law.[16] Newton went on to prove this mathematically, something Hooke was not able to do. For Hooke, the most hurtful aspect of the quarrel was the reluctance of some of his friends to take his side. While John Aubrey and others staunchly defended him, Wren seems to have remained silent, and Halley reported that Hooke and Sir John Hoskins, 'who till then were the most inseparable cronies', had fallen out and were hardly speaking.[17] They were soon meeting in coffee-houses again, but the sense of injustice rankled in Hooke.

If modern scholars have chewed over Hooke's contribution to clock and watch design, they have dined out on the extent to which he might (or might not) have been justified in claiming the theory of gravity. More recently, however, the emphasis has shifted. A recent assessment of the priority dispute with Newton concludes that a number of Hooke's contemporaries knew of the inverse square law, having reached it via different routes.[18] Instead of working according to a narrow and rather outdated view of scientific progress in which it seemed crucial to assign precedence to the single originator of a particular theory or invention, contemporary scholars are pursuing a more productive explanation of Hooke's and his contemporaries' working practices. Some have suggested that Hooke's use of diagrams and three-dimensional apparatus such as pendulums and rolling balls to simulate planetary motion, which to twentieth-century scholars seemed less persuasive than a mathematical demonstration, were in Hooke's day considered to be useful tools for exploring natural phenomena and were part of a broader concept of mathematics than our own. Hooke's mechanical and graphical explanations of his ideas, which fellows could understand even if they did not have mathematical training, reflected the early Royal Society's emphasis on public display and discussion.[19]

There had been a flurry of letters between Halley and Newton in the spring and summer of 1686. Halley explained that some time ago, he, Wren and Hooke had discussed the inverse square law; Hooke had claimed that he could demonstrate it, but failed to do so, saying that he would keep his demonstration until others had tried and failed. Now that Newton's own demonstration had been made public, Halley applied to Hooke again, but according to Halley, Hooke had stated that the inverse square law was 'but one small part of an excellent System of Nature, which he has conceived but has not yet compleatly made out, so that he thinks not fit to publish one part without the other'.[20] His system of nature was based on the principle of congruity, which governed the actions of the two 'Powers' of nature: matter and motion. Yet he never seems to have developed his ideas to the point where he was happy to publish them, and they remain scattered through his lectures and draft papers collected and published after his death.

Hooke's writings on the working of the brain may seem like a strange deviation from the scientific question he was trying to answer about the nature of light. And yet, despite their brevity and probably unfinished state, they explain Hooke's attitude towards the natural world, and man's attempts to understand it via the scientific method, in a way that accords perfectly with his other writings. Essentially Hooke believed that everything in nature followed the same physical laws, from the human brain to the solar system and beyond. This explains why he focused so strongly on 'similitudes', or analogies, between things that seemed on the surface to be very different. Knowledge of a tiny insect's anatomy, or an understanding of a physical process like joining a new axe head to a handle, could help to unlock the secrets of very different processes and anatomies, things that lay beyond the bounds of the human senses to explore directly. This trust in the universe's regularity lay at the heart of Hooke's working practice and explained his scientific method: he saw his method as an 'algebra' that, given the right data at the outset, would produce a correct result. His method enabled him to bring together a wide variety of information from across the fields he studied, all of which he believed could be explained by the physics of congruity.

EIGHT

A Discourse of Earthquakes

> Generation creates and Death destroys; Winter reduces what Summer produces: The Night refreshes what the Day has scorcht, and the Day cherishes what the Night benumb'd . . . All things almost circulate and have their Vicissitudes.[1]

It is unclear exactly when Hooke first wrote these meditative, almost poetic, words about the cycles of the universe, but he seems to have revised if not composed this passage in 1700. By this point in his life he was painfully familiar with vicissitude, or radical change. His beloved niece Grace had died in February 1687, an event about which we know little because it occurred during the gap in Hooke's diary-keeping. His friend and biographer Richard Waller believed the loss had affected Hooke profoundly, writing that Hooke never really recovered from the sorrow of Grace's death, and in addition was 'observ'd from that time to grow less active, more Melancholly and Cynical'.[2] The following year brought further change. In June a male heir had been born to James II, England's Catholic king, exacerbating fears among the Protestant populace that England was sliding towards Catholicism. A few months later James's Protestant son-in-law, William of Orange, was preparing to invade the country. On 5 November William's force landed in Torbay, and shortly after, James fled the country.

Almost six years after abandoning his previous diary, Hooke started a new daily journal on 1 November 1688. His brief entries reveal the fear and uncertainty Londoners were feeling during the period between the Dutch fleet's landing and the crowning of William and his wife, Mary, James's daughter, as joint monarchs in April 1689. He carried on with his day-to-day activities as usual, but for Hooke and his contemporaries who had lived through the English Civil Wars it must have revived difficult memories of a country turned upside-down.

Yet as the quote at the beginning of this chapter suggests, Hooke was no stranger to thoughts of vicissitude. The notion of a changing earth – one that had experienced tumultuous events that replaced oceans with land and land with seas – was not new in the seventeenth century, but it was one that received growing attention and debate among scientists and theologians.[3] It was a topic that intrigued Hooke throughout his life, and one that occupied him particularly in the 1680s and '90s. When editing Hooke's posthumous works, Richard Waller printed almost two hundred pages of Hooke's writings under the heading 'A Discourse of Earthquakes', but, as we will see, this research drew together a number of Hooke's interests: earthquakes, volcanoes and fossils, but also astronomy, physics and microscopy, foreign travel and languages and even classical mythology and antiquarian studies. In this long-running series of lectures Hooke argued that Earth was much older than contemporary accounts held it to be. It had undergone a long series of changes caused by earthquakes and volcanoes, so that what was once at the bottom of the sea was now at the tops of mountains. He suggested that England had once stood in the 'Torrid Zone' covered by tropical seas, with fish swimming over what were now green fields.[4] Fossils were the remains of living creatures, and some of these species were now extinct.

These were radical and unorthodox ideas, and before investigating them in more detail we need to ask how Hooke arrived at them. In order to build up this new understanding of Earth's history, he first had to break apart old notions. The fact that he was willing to do this shows his commitment to a particular aspect of his research method. It was of the greatest importance, he wrote, that philosophers should proceed with an open mind, unswayed 'by this or that Opinion' and free from prejudice. They would only progress by focusing on the research goal, ignoring the distractions of amusing or lucrative experiments and paying attention to the small signs that indicated the right path, just as Columbus noted clouds, seaweed and shallower seas that guided him on his course. This may sound straightforward, but Hooke realized it was easier said than done. In fact, human nature, or what Hooke called 'the Constitution and Powers of the Soul', fundamentally shaped the production of knowledge, with its distractions, prejudices and deference to ancient opinions. In order for knowledge to be increased, human nature itself had to be dissected and rectified.[5]

Hooke's discussion of 'the Perfections and Imperfections of Humane Nature' came at the beginning of his philosophical method, signalling its importance. In all this material he was strongly influenced by Francis Bacon, who decades before had suggested that humans were all prone to certain stumbling blocks, or 'idols', that stood in the way of knowledge. These idols were caused by human nature and individual prejudices as well as systems of communication and beliefs. Following Bacon, Hooke argued that no progress could be made by a mind unprepared for doing science. The problem was that human minds and bodies were not made for science. Hooke argued that our understanding of the world around us, provided by our senses of sight, taste and so on, is suited to our own species, and that 'if there were another Species of Intelligent Creatures in the World, they might have quite another kind of Apprehension of the same thing'. Furthermore, humans should not prioritize their own species when it came to understanding how the natural world worked: 'We ought to conceive of things as they are part of, and Actors or Patients in the Universe, and not only as they have this or that peculiar Relation or Influence on our own Senses or selves.' Today's scientists might see these ideas as part of a movement towards scientific objectivity, a value that we now believe should underpin reliable scientific research. But for Hooke's contemporaries, who believed that God's special providence had arranged the whole universe for mankind's enjoyment, this would have been a difficult pill to swallow.[6]

Other problems stemmed from human psychology. Again following Bacon, Hooke argued that everyone's 'own peculiar Structure' (of mind and body) affects our understanding: 'a melancholy Person, that thinks he meets with nothing but frightful Apparitions, does convert all things he either sees or hears into dreadful Representations, and makes use of them to strengthen his Phant'sy, and fill it fuller of such uneasy Apprehensions.' Similarly, people incline to their own interests: the whole philosophy of a chemist is tinged with chemical ideas, and astrologers try to show that everything operates under the influence of celestial bodies. Other prejudices stem from a person's individual experience, from 'Language, Education, Breeding, Conversation, Instruction, Study, from an Esteem of Authors, Tutors, Masters, Antiquity, Novelty, Fashions, Customs, or the like'. Our existing languages are not philosophical: new words are needed to express ideas more clearly and precisely, and other words need to be 'blotted out' because they signify things that do not exist. Education leads to some ideas being taken for granted and never questioned, and similarly, we find it hard to question the beliefs of an author or teacher whom we admire. But, Hooke warned, we need to guard against admitting erroneous beliefs into our minds because admitting one wrong idea will open the door to many others: 'Error being a kind of Ferment which tends to the turning or conforming all things to its own Nature, and like an infected Person has Influence on all things it comes near.'[7]

For Hooke, the most frustrating difficulty was what he called 'Prejudice'. Prejudice existed in two main forms: the inclination to agree with what had been written by ancient writers; and the tendency to accept theories that supported one's own existing theory of the world. Thus Hooke suggested two main solutions. When weighing up other people's ideas, a philosopher should not think about the person making the statement but 'how true the things are [that] he asserts'. Here Hooke seems to be deliberately opposing what some historians have suggested was the Royal Society's practice of assenting to reports provided by gentlemen or noblemen solely on the grounds of their social status. His second solution to prejudice encouraged what he called 'an Hypothetical Scepticism', a radical programme of mental adjustment 'whereby to impose upon our selves a Disbelief of every thing whatsoever, that we have already imbraced or taken in as a Truth'. Truths should only be readmitted after 'farther Tryals or Experience' had shown them to be fact.[8]

Here Hooke was following other great thinkers of his age: Francis Bacon and René Descartes had both stressed the need to reassess received opinions as a first step in reforming knowledge. 'Knowledge is made by oblivion,' wrote Sir Thomas Browne, in his characteristically epigrammatic style.[9] But it was not easy to assign commonly held opinions to oblivion and start all over again, and most philosophers were only willing to go so far in their scepticism. This was particularly the case when the facts were drawn from the Bible, and therefore assumed to be endorsed by God.

Hooke seems never to have expressed any doubts about the existence of God. We can assume that, like his contemporaries, he attended religious worship regularly, although his diary does not mention many church services, so perhaps he did not find them as stimulating as others did. Richard Waller pointed out that in his diary entries Hooke always acknowledged God on auspicious occasions or after making a new discovery, but he somewhat spoiled the effect by adding the obscure comment 'If he was particular in some Matters, let us leave him to the searcher of Hearts.'[10] Perhaps Hooke had shared the heterodox opinions of his coffee-house companion Francis Lodwick, who believed that there had been men on earth before Adam. Other Royal Society colleagues were famously devout: Robert Boyle worked hard in his own writings to reassure his readers that the new philosophy presented no threat to established religion, as some worried that it might.

Hooke was seemingly not particularly concerned about this possibility, but as we might expect, seems to have come closest to his Maker when viewing the humblest parts of creation through a microscope. In particular, the fact that insects grew from invisibly tiny eggs and underwent such unexpected metamorphoses during their lives provided what he called 'a very coercive argument to admire the goodness and providence of the infinitely wise Creator in his most excellent contrivances and dispensations'.[11] *Micrographia* is punctuated with similar statements. For Hooke's contemporaries, God had given mankind two ways of learning about Him and his works: his words, in the Bible; and his creation, often called the 'book of nature'. In Hooke's eyes, however, the book of nature came first. If his experiments and observations threw up results that contradicted the biblical account, then he was willing to question God's word as expressed in the Bible. In 1689 he read a paper at a meeting of the Royal Society about 'the theory of the Earth'. The minutes of the meeting record some of the content of his lecture:

> that the design of the Scriptures being not so much to teach Naturall Philosophy, as to shew the wisdom and power of God in making the World as it is, what contributed to this perswasion, was the Theory most proper to be delivered in Scripture. Accordingly [Hooke argued] that the explication of the Creation was adapted to the most common apprehensions of Mankind, and that all the expression of holy-writt seemed to be so worded.[12]

That is, Hooke argued that since the point of the biblical creation story was to show God's wisdom and power (not to teach men about natural history), the ancient writer of the book of Genesis had adapted his story for that goal, and he had used language suited to the way that the average reader understood the world, not the language of philosophers. The creation story was not intended to be taken literally as a philosophical account of the beginning of the world.

As radical as it might sound, Hooke was not the first to make this suggestion, and while it was controversial, it was not dangerously so. Hooke's friend Edmond Halley also argued that the biblical account was unsatisfactory, although for different reasons.[13] For Hooke, however, his point regarding the creation story was part of a long-running series of speculations about the earth and its inhabitants, and a geological theory that was unlike anything his colleagues had suggested previously. His willingness to jettison the Bible on the grounds that it did not agree with the evidence of natural history is consistent with his warnings against prejudice and preconceived opinions. But in order to construct his new theory of the Earth, he first needed to draw together material from many sources.

Despite hardly ever leaving London physically, Hooke spent a good deal of his mental time travelling the world via books, maps and conversations with travellers. There were several reasons why he was addicted to armchair travel. In terms of his scientific methodology it seemed fundamental to him that observations and artefacts were collected from as wide a field as possible. 'Histories and Observations from abroad' were needed alongside local data: he suggested that in order to investigate the human body, it would be necessary to compare, for example, a 'wild Irish Man' with people from the Cape of Good Hope and elsewhere to find out what was innate and what was shaped by local conditions. Hooke was not alone in his fascination with foreign travel, and this had been a significant area of investigation for the Royal Society since its foundation. Robert Boyle had produced a long questionnaire for those travelling to foreign parts, and the society sent copies to those they thought might be able to help. Other fellows were asked to draw up sets of questions relevant to specific places, such as Greenland. The response rate was a little disappointing for the society, but sometimes, when a traveller had more time or enthusiasm, a mass of information was sent back to London. Some of these questions seem naive now – such as whether diamonds grew again in places where they had been extracted – but they reflect the state of knowledge at the time, and present a vivid picture of the thirst for information supplied by credible first-hand witnesses. Many of the existing books describing foreign parts were a mixture of fact and fiction, and it was hard to distinguish between the two. Following a discussion about the diet of reindeer in July 1693, the minutes of the Royal Society note that 'it was upon this occasion observed that Divers french Authors . . . had been of late yeares published, which were filled with Fables, and Falsitys. As the History of the Antilles, of old and new Athens, of Tunquin, of a Voyage into Africa, &c.'[14]

This helps to explain Hooke's enthusiasm for new travel accounts, particularly those written by trusted eyewitnesses. As we have seen, he supported Robert Knox's description of Ceylon through the press. He also encouraged the East India Company employee Samuel Baron to write a description of the kingdom of 'Tonqueen', or present-day northern Vietnam. Baron, whose father was European and mother Vietnamese, had been born in Tonqueen and was employed by the English East India Company. He had met Hooke in London in 1677 and Hooke had kept in touch, encouraging Baron to correct the errors in Jean-Baptiste Tavernier's account of Vietnam, which had been published in Paris in 1676 (probably the history of 'Tunquin' mentioned above). Baron sent his manuscript to Hooke and Sir John Hoskins in 1686. The original manuscript is now lost, but the text was printed in the early eighteenth century, and Baron's lavish set of illustrations still survives in the archives of the Royal Society (illus. 26). Hooke was also involved in his friend and Royal Society colleague Alexander Pitfeild's translation of a new book about Siam (Thailand), which had been published in French by Simon de la Loubère. The *New Historical Relation of the Kingdom of Siam* included material that would have fascinated Hooke, particularly sections about the Thai language, Chinese chronology and Indian mathematics. There is a gap in Hooke's diary for this period, but the remaining text shows that he was liaising between Pitfeild and the publisher, Thomas Horne, while the book was in production.

Part of the appeal of texts like these stemmed from economic considerations. As Baron remarked about his own account of Tonqueen, the narrative provided information 'sufficient for a New commer, to Conduct buseness by, att his first Entrance' into Vietnam. Accounts of foreign parts were expected to provide an insight into local manufacturing, agriculture and natural resources with enough explanation of local customs to potentially smooth the way for a mercantile relationship. As we have seen, Hooke was always curious about trades, and he helped innovators such as John Dwight and Patrick Barrett explore

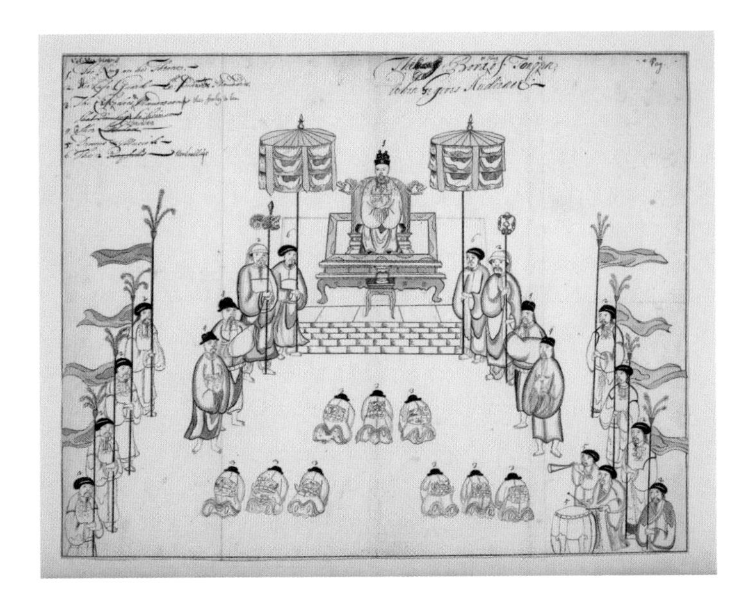

ways to mimic imported goods. Occasionally he also tried this himself. His attempt to recreate what he called in his diary 'Chickee linnen' (perhaps Cherokee?) is recorded in the minutes of a Royal Society meeting held in December 1689:

> [Hooke] described a Sort of Cloath, which, He said, was made by the Indians in America about 500 Miles [805 km] to the Northwest of Carolina, being a course sort made of a Substance, that is very strong, but neither flax, nor hemp, nor like any Materiall, he had hitherto seen in England. But he shewed the manner of the Texture thereof, which he imitated by bobbins, as they weave bone-lace.[15]

In this case Hooke was not interested in manufacturing the new type of cloth for profit, but instead imitating it as a way of understanding how it was made.

26 Illustration of the 'Bora or King of Tonqueen', accompanying Samuel Baron's description of Tonqueen, *c.* 1685.

Hooke made his own small contribution to the growing literature about foreign places by translating two letters written by French Jesuit priests in China and publishing an article of his own on the Chinese language. As a vast and literate civilization with its own traditions of science and medicine (not to mention manufacturing) that was almost entirely closed to Westerners in the period, China was probably the most fascinating of all foreign countries to British philosophers. The Royal Society regularly speculated on Chinese topics. At one meeting, Hooke demonstrated the process of moxibustion (burning a small amount of moxa, a preparation of dried mugwort, next to the skin), a traditional Chinese therapy that was supposed to cure gout. At another meeting Sir Hans Sloane presented the society with a gift of several Chinese surgical instruments and ear-cleaning tools, together with a statuette depicting their use (illus. 27). In 1679 Hooke himself had called on a Chinese

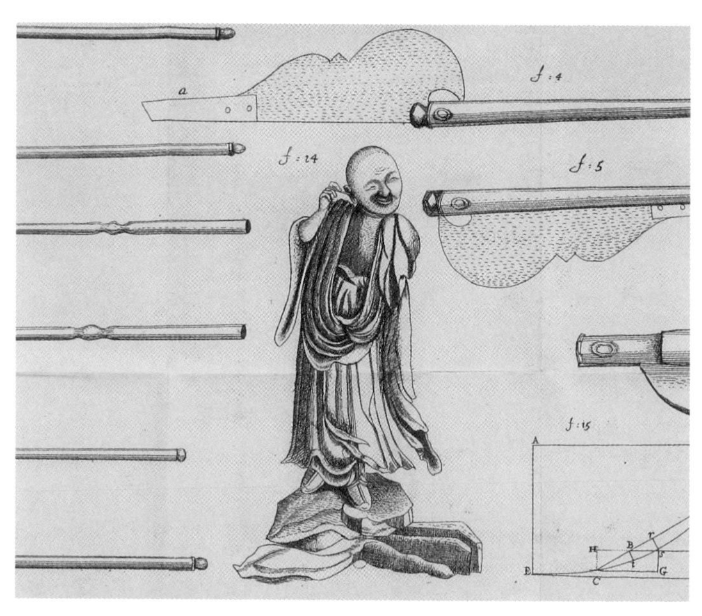

27 A Chinese statuette depicting the use of an ear-cleaning tool, illustrated in *Philosophical Transactions of the Royal Society* (1698).

merchant, possibly visiting London in connection with the East India Company, the trading tentacles of which extended throughout East Asia. The next day Hooke was given a pair of chopsticks by Charles Chamberlain, an East India Company agent. Ten years later he noted in his diary that he had taken tea with two 'Bantam ladies' at his friend Mrs Moore's: Bantam was a major East India Company trading post in Indonesia, but it was rare to hear of women visiting London from such far-flung places during Hooke's lifetime.

Hooke's article about the Chinese language stemmed from his interest in artificial languages, which he shared with some of his close friends in the Royal Society. His early mentor John Wilkins had proposed a new philosophical language in the 1660s, intended to put an end to misunderstandings caused by slippery words and to enable communication between people from different countries. Hooke, Francis Lodwick and others continued Wilkins's work after his death. Hooke was particularly interested in Mandarin because he thought it was just such an artificial language. Chinese had been discussed by Western writers for some time, and Hooke was up to date with the latest theories. With Francis Lodwick, he had tea with some Chinese visitors to London in July 1693, but after the visit he recorded in his diary, 'I could learn little: 8 or 10 characters pronouncd all alike but of Differing signification.'[16] In his article he argued that the words used to pronounce the Chinese characters seemed to be 'Arbitrarily Imposed' on the characters, just as the different European languages all impose their own words on a shared set of numerals.[17] Hooke studied the Chinese characters closely, with the help of the few examples of Chinese texts, dictionaries and commentaries that he could get his hands on. He concluded that the original Chinese language that used these characters had been lost, and, radically, he suggested that the original pronunciation of other ancient languages such as Hebrew, Syriac, Arabic,

Greek and Latin might also have been completely different from the current one.[18] All things have their vicissitudes, even the languages of the Bible and the great works of the Western literary canon.

Hooke continued to search out accounts of travel to China. In a lecture to the Royal Society in 1689, he discussed his translation of a recent German book about the journey of a Russian ambassador, Fyodor Isakovich Baykov, from Russia overland to China in 1658. Only a partial draft of the lecture remains among Hooke's papers, but it shows that he tried to compare the path travelled by Baykov in his memoir with the most up-to-date map of the region. He complained that it was impossible to know the true location of places because the description of Baykov's voyage did not record 'Latitude nor longitude nor Distance, noe nor what way or quarter one place lyes from the other':

> Every one of which notwithstanding might with very little trouble have been observed and Recorded. And thereby it would have been easy to have placed all the way or Rode with the townes Residences Hills Rivers etc in the Due Situation, but this It seems was not to be expected from a Russe Embassador.[19]

Hooke's exasperation here stemmed from his conviction that it should be straightforward to record an overland route accurately – or easier, at least, than plotting the course of a ship at sea, which was routinely recorded in the ship's log. Hooke argued that it could be done 'by a very small apparatus of Instruments and with a very little trouble if Some person of the Retinue were furnished with such an apparatus and some few Rules of Instruction, and were obleigd to Make the Necessary observations and keep a Journall of them.'[20] As always for Hooke, the problem could be solved with instruments. He had himself invented devices intended for travellers, mostly seafarers: an instrument for sounding the depth of the sea using a float rather than a line, which at the same time gathered samples of seawater; his marine barometer; inventions for finding the longitude, including his improved watches, an instrument for measuring the distance between the moon and stars and an improved backstaff (used for measuring the sun's altitude); and a 'way-wiser', which would measure the distance and direction of a ship's travel.[21] Travellers who used such equipment were a vital source of experimental data. Hooke mentioned that he had 'Engaged' his friend Robert Knox to investigate the 'differing motion of the Pendulum in Differing Latitudes' during Knox's voyages to and from the Indies in the early 1690s, having bought him apparatus and instructed him on how to use it.[22]

All this activity notwithstanding, Hooke's fascination with accounts of far-flung places mostly stemmed from his enduring interest in the earth itself. His series of lectures on 'earthquakes' were composed and delivered from the 1660s right through to the 1690s, and Hooke was still adding material as late as 1700. They wove together the reports Hooke had gathered, and as we have seen they came to some remarkable conclusions that challenged orthodox thinking at the time. As a boy Hooke had seen shells embedded in high cliffs on the Isle of Wight, far above the current sea-level. Throughout his life he collected stories of earthquakes and volcanic eruptions, and of visits to deep wells, caves and mountain peaks (such as the 'Pike' in Tenerife, a famously vertiginous summit in the period), places where vapours issued from the ground and petrifying springs. He also collected the strangely shaped stones that we now know to be fossils (illus. 28). Although some earlier thinkers had suggested an organic origin for fossils, the prevailing opinion was that they were 'figured stones', produced naturally as a kind of joke by nature in imitation of living plants and animals. Hooke dismissed this idea:

> it seems very difficult to imagine that Nature formed all these curious Bodies for no other End, than only to play the Mimick in the Mineral Kingdom, and only to imitate what she had done for some more noble End, and in a greater Perfection in the Vegetable and Animal Kingdoms.[23]

He had inspected fossilized shells with his microscope and observed the same structures that he had seen in living creatures. He argued that these petrified shells found all over the world were, in fact, either the remains of living creatures or the impressions left by them in what was now rock but had once been mud or even water. To counter the argument that many of the fossils (huge ammonites, for example) seemed to represent creatures that did not exist, he further argued for radical species change:

> we will, for the present, take this Supposition to be real and true, that there have been in former times of the World, divers Species of Creatures, that are now quite lost, and no more of them surviving upon any part of the Earth. Again, That there are now divers Species of Creatures which never exceed at present a certain Magniture, which yet, in former Ages of the World, were usually of a much greater and Gygantick Standard . . . we will grant also a supposition that several Species may really not have been created of the very Shapes they now are of, but that they have changed in great part their Shape, as well as dwindled and degenerated into a dwarfish Progeny; that this may have been so considerable, as that if we could have seen both together, we should not have judged them of the same Species.[24]

28 Ammonite fossils engraved from drawings by Hooke, printed in Hooke, *Posthumous Works* (1705).

In other words (and as he put it elsewhere), it was very likely that some species had been lost, and that others living now had not been present 'from the beginning'.[25]

This was one of Hooke's most unorthodox theories. Most of his contemporaries would never have questioned the idea that the animals around them had remained the same since Noah had rescued their distant ancestors from the biblical flood in his ark. Why would God have bothered to ask Noah to save them if they were destined to die out anyway? But as we have seen, Hooke put the biblical account to one side and focused on what other evidence suggested. This was where his observations, data and stories collected from around the world became significant. He drew a connection between the formation of what we would now call stalactites and stalagmites, which he knew from eyewitness reports, and the way that sediment could harden into rock. He combed through ancient writers to find references to geological events, suggesting that stories in classical literature about giants that waged war on the gods by throwing 'huge Stones and Fire at Heaven' were simply poetical descriptions of earthquakes.[26]

Hooke spent a long time trying to convince his colleagues that fossils really were organic in origin, without much success. His theories on fossils supported his belief in an unstable and ever-changing earth. As we have seen, he argued that over the ages mountains and oceans had changed places. He also discussed the Earth's shape, agreeing with Newton that Earth was slightly oval-shaped, or as Hooke put it, 'Turnep-form', wider at the equator than the poles. He suggested that the planet's centre of gravity might change over time, and that the 'Magnetical Poles and Meridians of the Earth' had also shifted. Time changed everything – 'we may even doubt whether the power of Gravity itself may not alter in time' – but even time itself might change. He suggested that Earth, which had been set spinning at creation like a top, might over time spin less quickly. Therefore the long lives recorded for the descendants of Adam (for example, Methuselah, who died aged 969 years) might actually have been no longer than men's lives ordinarily are now: 'for though perhaps they might number more Revolutions of the Sun, or more Years than we can now, yet our few Years may comprehend as great a space of time'.[27] Since the date of the world's creation, famously calculated by Archbishop James Ussher as 23 October 4004 BC, relied in part on adding together the lifespans of Adam's descendants, this was an unsettling idea.

In some ways Hooke's lectures on earthquakes provide a kind of masterclass in his philosophical method. The texts were not prepared by Hooke for publication, but rather were gathered together by Richard Waller from Hooke's papers after his death, and it seems likely that they were printed in much the same state as they might have been delivered orally at a series of Royal Society meetings. As such, they give us a glimpse of what it might have been like to listen to Hooke in person, trying to convince his colleagues of the truth of his proposals about the causes of fossils, the mechanisms of fossilization and the transformations that had taken place in the earth. His different lectures focused on different aspects of the subject, but he always took care to present his evidence clearly and persuasively. In one of his lectures he offered nine separate arguments to support his proposition that it was possible for water to be transformed into a durable, 'stony Substance'.[28] This proposition was itself part of a set of arguments intended to show that fossils were the remains of living things: if water could congeal into stone, then it would be easier to believe that animals or plants could be fossilized by 'having their Pores fill'd up with some petrifying liquid Substance'.[29] We may not agree with the details of Hooke's evidence, but his method is instructive.

His nine arguments to support this proposition about water used all the tools at Hooke's command. As always, he reached first for his microscope. He noted that almost all streams and running waters contained a kind of sand that, when inspected through the microscope, was discovered to be 'curiously wrought and figured' like crystal or diamond. He was familiar with all sorts of crystalline structures, and all of them were originally liquid, so he argued that the water was producing these crystal sands. His second point relied on his observation of trade practices. He argued that the process of making salt in a saltern led to 'great quantities of Sand' seemingly 'coagulated' out during the boiling of the seawater, which suggested that there was something soluble in the water that turned solid. His next point drew on expert testimony: 'an eminent Physician' had assured him that distilling the same water several times turns it into a white 'Calx', which is insoluble. His fourth argument referred to the kind of eyewitness accounts he collected so assiduously: in the caves in the Peak District in Derbyshire, stalagmites and stalactites were generated by water trickling through the roof. His fifth, sixth and seventh points all drew on evidence from printed books, and his final two arguments referred to experimental proofs (although his ninth point simply says that he could provide some experiments to illustrate his points but felt that the weight of the previous points should suffice).[30]

Here we can see Hooke not just trying to convince his audience that living things could actually be turned into stone through the action of water but demonstrating the process of reasoning that he himself followed. None of these instances was particularly persuasive alone, but together Hooke felt they were like the hints and signs that led Columbus towards the New World.

Epilogue: The Teeth of Time

Hooke remained convinced that he was right about fossils, but his later lectures suggest that his colleagues were not persuaded. In May 1689 he began yet another defence of his theory:

> I Delivered in my last Lecture in this place, the Methods I had made use of [to establish his theory explaining the current state of the earth] . . . But notwithstanding all the Arguments I have alleged, and the Proofs I have produced in the delivery of this Theory, I still find that there remain upon the Minds of some such Doubts and contrary Persuasions, that they cannot forsake their former Opinions.

Despite having built up his argument by 'a methodical Induction' from the data, Hooke had been unable to persuade his audience to take the step that he had taken such a long time before and rid their minds of biases and prejudices. He rather wearily explained that he would in this present lecture answer again some of the main objections to his theory. In doing so, he hoped that

> the *Idola* [Bacon's idols] which pre-possess the Minds of some Men, and molest them in the discovery and imbracing of Sciences may be detected, and, as much as may be,

> removed and dissolved, thereby to leave the Mind more free to Discourse and Reason aright, without the prejudices of any unsound, unaccountable and unwarrantable Doctrines formerly imbrac'd.[1]

Baconian to the last, Hooke still believed that if only their minds were free to think clearly, his colleagues would see what he could see.

Hooke continued to read lectures at Gresham College and Royal Society meetings in the 1690s, revising his older material on earthquakes and fossils when new evidence became available. In December 1691 he was awarded the honorary degree of Doctor of Medicine by his old friend John Tillotson, Archbishop of Canterbury, in recognition of his many contributions to knowledge. He was still in demand as a surveyor and overseer of building works: he completed a grand hospital, or almshouse, for the Company of Haberdashers in 1695. In July 1696, on the very day of his 61st birthday, Hooke's court battle with the estate of Sir John Cutler was finally settled in his favour. Cutler had died in 1693 without ever resuming his promised payments to Hooke for the Cutlerian lectures, and the settlement felt like 'a new Birth' to Hooke.[2] But like the watch mechanism that he had described so vividly, he was finally slowing down. Richard Waller painted a rather gloomy picture of his final years. Hooke had planned to repeat many of his former experiments and 'finish the Accounts, Observations and Deductions from them', and to produce descriptions of all the instruments he had invented, but he was prevented, Waller says, by increasing ill-health. Having been very active in his youth and middle age, from 1697 (when Hooke was 62) he began to complain of swelling and pain in his legs, breathlessness and giddiness. His sight deteriorated towards the end of his life, so much so that he could not see to read or write. Practically bed-ridden in his last year, he continued in what Waller called a 'dying Life' for some time, until death finally released him on 3 March 1703. He was buried in the church of St Helen's, Bishopsgate, where his niece Grace had been buried, attended by his Royal Society colleagues and friends.

Even after his death, Hooke continued to exert an influence over the Royal Society. Waller managed to retrieve the bulk of Hooke's own notes and papers from his executors, and from time to time he read out some of this material at meetings of the society. Hearing about Hooke's unpublished method for improving philosophy, and the lectures on light, comets, earthquakes and so on, the society asked Waller to edit Hooke's posthumous works in November 1703. The edition was published in 1705, with a dedication 'to Sir Isaac Newton, K[nigh]t, President, and to the Council and fellows of the Royal Society of London'. Waller's *Posthumous Works* did not include all of Hooke's archive, however, and in February 1710 the Royal Society council 'earnestly desired' Waller to publish a further collection from the remaining papers. Nothing came of this idea, but more than a decade after Hooke's death Waller was still reading new material at society meetings, including accounts of inventions and travels in China and elsewhere: in June 1714 he presented a description of the 'wonderful Palme-tree of Ceilon' that had been found among Hooke's papers.[3] After Waller's own death in 1715, Hooke's remaining papers were passed to William Derham, another fellow of the Royal Society. Derham had only been elected to the fellowship in 1703, and he had not known Hooke as well as Waller had done, but he published a further, more miscellaneous volume of Hooke's papers in 1726.

Newton had been elected president of the Royal Society in November 1703. It has often been suggested that Newton waited until after Hooke's death to accept the presidency; it is certainly possible, but there is no evidence that this was the case. Newton had already served on the society's council alongside Hooke in

1697 and 1699 (although this may not have involved much activity or indeed contact), and there was at least one other active fellow after 1703 with whom Newton had quarrelled bitterly, the astronomer royal John Flamsteed. The story that Newton destroyed a portrait of Hooke hanging at the Royal Society is sheer fantasy, and there is no strong evidence to suggest that such a portrait even existed. It is clear from the activities of Waller and Derham that, rather than trying to erase his legacy, in the decades after his death Hooke's colleagues saw him as a respected figure whose life and work were worth celebrating in print. But the memory of his work faded during the eighteenth century, after those who knew him had died. There is little evidence that Hooke's philosophical method influenced his peers at the Royal Society. He certainly attempted to bring them around to his point of view, sometimes commenting about the ideal scientific practice in his papers read at the society, or as we have seen, giving step-by-step examples of his method.[4] The more mathematical, or Newtonian, approach to investigating the natural world that dominated the society in the eighteenth century must have made Hooke's emphasis on fact-finding, collaboration and the psychology of the observer seem quaintly out of date. In *Micrographia* Hooke had described a silverfish, which he had caught in the act of busily chewing through his books and papers, as 'one of the teeth of Time'.[5] Had this uncharacteristically poetical interruption been occasioned by thoughts of his own oblivion?

And yet perhaps it is worthwhile to consider what Hooke's method can tell us about the challenges of doing scientific research, particularly at a time when its public audience was sceptical or downright hostile. Hooke invited his readers to join him in his observations of the natural world. He emphasized participation above all: although they would be useful, no particular skills or equipment, and no training in mathematics, were absolutely necessary; nothing except sincerity. He dwelt thoughtfully on the role of the observer, and in particular on how his or her physical body and mental prejudices influenced the production of knowledge. He expressed this in different ways in his own work: for example, writing himself into *Micrographia* in the character of narrator and guide, hoping that introducing himself to his audience would build a relationship of trust between scientist and reader. His method was collaborative: he constantly looked outward from his own research to the activities of people from all social spheres, valuing (or, some might say, appropriating) their individual expertise. Above all, he communicated his sense of optimism in scientific progress. At a Royal Society meeting in 1675, his colleague the anatomist Dr William Croone read a lecture about flight in birds, concluding that human flight was impossible. 'Hereupon', Hooke noted in his own record of the meeting, 'RH suggested that what nature Did not supply art and Reason could.'[6] Hooke brought together 'art', or mechanical skill, with reason, in a way that few of his contemporaries could match.

CHRONOLOGY

1635	Born on 18 July in Freshwater, Isle of Wight
c. 1648	Moves to London after the death of his father, John; begins studying with Richard Busby, head of Westminster School
c. 1653	Enters Christ Church College, Oxford
c. 1657	Begins working as Robert Boyle's assistant in Oxford
1660	Charles II crowned as king; the Royal Society is founded
1661	Publishes his first surviving work, *An Attempt for the Explication of the Phaenomena Observable in an Experiment Published by the Honourable Robert Boyle, Esq*
1662	Begins working for the Royal Society as 'curator of experiments'
1664	Cutlerian lectureship founded; Hooke moves into Gresham College
1665	*Micrographia* is published; Hooke is elected as Gresham Professor of Geometry
1666	The Great Fire destroys much of London
1667	Appointed as city surveyor
1670	Commissioned to build premises for the Royal College of Physicians, including an anatomy theatre (completed by 1679)
1672	Begins writing his 'memoranda', or diary
1674	In March, publishes *An Attempt to Prove the Motion of the Earth from Observations*, the first Cutlerian lecture to be published; in April, starts work on Bethlem Hospital (completed in 1676); in September, begins work on Montagu House, Bloomsbury (completed in 1680); in December, publishes *Animadversions on the First Part of the Machina coelestis*

1675	Christiaan Huygens's spring-regulated watch is demonstrated at the Royal Society, sparking a dispute with Hooke. In October, Hooke publishes *A Description of Helioscopes, and Some Other Instruments*
1676	In September, publishes *Lampas: or, Descriptions of Some Mechanical Improvements of Lamps and Waterpoises* (dated 1677)
1677	Henry Oldenburg dies; Hooke is elected one of two secretaries to the Royal Society
1678	In March, publishes *Lectures and Collections*; in November, publishes *Lectures de potentia restitutiva; or, Of Spring*
1679	The first volume of Hooke's journal, *Philosophical Collections*, appears; in total seven volumes are published between 1679 and 1682
1687	Delivers a series of lectures at the Royal Society on geology and geomorphology (continued into the 1690s)
1688	Begins writing a new diary
1691	Receives doctorate in medicine from the Archbishop of Canterbury, in recognition of his research
1696	Lawsuit with Sir John Cutler's estate over salary arrears for Cutlerian lectures is settled in Hooke's favour
1703	On 3 March dies in his rooms at Gresham College

REFERENCES

Hooke's diary entries have been cited in this book as transcribed from his original diary manuscripts, now London Metropolitan Archives CLC/495/MS01758 and British Library Sloane MS 4024. However, for ease of access, they are referenced here in modern editions.

Introduction: Mad, Foolish and Phantastick

1 Robert Hooke, *The Posthumous Works of Robert Hooke*, ed. Richard Waller (London, 1705), p. 39.
2 Ibid., pp. 6, 7.
3 All these books are highly recommended: Francesco G. Sacco's *Real, Mechanical, Experimental: Robert Hooke's Natural Philosophy* (Cham, 2020) provides a scholarly assessment of Hooke's scientific programme; Lisa Jardine's *The Curious Life of Robert Hooke: The Man Who Measured London* (London, 2004) and Stephen Inwood's *The Man Who Knew Too Much: The Strange and Inventive Life of Robert Hooke, 1635–1703* (London, 2002) are biographies aimed at a more general readership.

1 The Present Deficiency of Natural Philosophy

1 Robert Hooke, *The Posthumous Works of Robert Hooke*, ed. Richard Waller (London, 1705), p. 3.
2 Francis Bacon, *The Oxford Francis Bacon*, vol. IV: *The Advancement of Learning*, ed. Michael Kiernan (Oxford, 2000), p. 32.
3 Hooke, *Posthumous Works*, p. 3.
4 Richard Waller, 'The Life of Dr Robert Hooke', in Hooke, *Posthumous Works*, p. ii.
5 Ibid., p. iii.

6 Francesco G. Sacco, *Real, Mechanical, Experimental: Robert Hooke's Natural Philosophy* (Cham, 2020), pp. 189–91.
7 Thomas Birch, *The History of the Royal Society of London* (London, 1756–7), vol. I, pp. 123–4.
8 Ibid., vol. I, p. 250.
9 Ibid., vol. I, p. 26.
10 Ibid., vol. III, p. 425.
11 Hooke, *Posthumous Works*, p. 18.
12 'Five Ways to Compute the Relative Value of a UK Pound Amount, 1270 to Present', www.measuringworth.com, 8 June 2023.
13 Birch, *History*, vol. I, p. 503.
14 Hooke, *Posthumous Works*, p. 7.
15 Ibid., pp. 18–21.
16 Ibid., pp. 20–21.
17 Ibid., pp. 21–4.
18 Ibid., pp. 33, 36.
19 Ibid., pp. 36, 37.
20 Ibid., p. 42.
21 Ibid., p. 34.

2 A City Where All the Noises and Business in the World Do Meet

1 John Evelyn, *The Diary of John Evelyn*, ed. E. S. de Beer (Oxford, 1955), vol. III, p. 416.
2 Robert Hooke, *The Posthumous Works of Robert Hooke*, ed. Richard Waller (London, 1705), p. 21.
3 Robert Hooke, *The Diary of Robert Hooke*, ed. Henry W. Robinson and Walter Adams (London, 1935), p. 357 (8 May 1678); Felicity Henderson, 'Unpublished Material from the Memorandum Book of Robert Hooke, Guildhall Library MS 1758', *Notes and Records of the Royal Society*, LXI/2 (2007), pp. 129–75 (p. 154: 11 August 1682).
4 Hooke, *Diary*, p. 251 (28 September 1676). In 1661 Hooke's Royal Society colleague John Evelyn had published *Fumifugium; or, The inconveniencie of the aer and smoak of London dissipated. Together with some remedies humbly proposed* (London, 1661), drawing attention to the problem of London's air pollution.
5 Hooke, *Diary*, p. 234 (26 May 1676).
6 Richard Waller gave the only description of Hooke's plan in his biography of Hooke (Richard Waller, 'The Life of Dr Robert Hooke',

in Hooke, *Posthumous Works*, pp. xii–xiii), although this was only from hearsay. The plan itself seems to have been lost, but it has previously been identified with one depicted in a contemporary engraving printed in Amsterdam by Marcus Willemsz Doornick ('Nieuw modell om de afgebrande Stadt London te Herbouwen', 1666), seemingly solely on the grounds that this plan is also in a grid format.

7 An excellent and very detailed description of Hooke's surveying work for the city can be found in Michael Cooper's *Robert Hooke and the Rebuilding of London* (Stroud, 2003).

8 Hooke, *Diary*, pp. 406 (10 April 1679), 293 (30 May 1677; where 'Drunken' has been mistranscribed as 'Brounker').

9 London Metropolitan Archives, COL/CC/CLC/07/002, f. 96v, report dated 24 January 1674.

10 Hooke, *Diary*, p. 79 (9 January 1674).

11 John Conyers, 'A Letter of Mr. John Conyers . . . in Which Letter is Contained a Draught and Description of a Very Useful and Cheap Pump', *Philosophical Transactions*, XII/136 (1677), pp. 887–90 (pp. 888–90); Hooke, *Diary*, p. 82 (22 January 1674); Hooke, *Posthumous Works*, p. 335.

12 Evelyn, *Diary*, vol. IV, p. 344.

13 There is some confusion about the exact dimensions of Hooke's Bethlem Hospital, but in 1720 John Strype described the length as 'from *East* to *West* 540 foot' (Strype, *A Survey of the Cities of London and Westminster* (London, 1720), vol. I, p. 192).

14 Hooke, *Diary*, p. 428 (22 October 1679).

15 Ibid., p. 126 (14 October 1674).

16 London Metropolitan Archives, COL/SJ/16/003, order dated 20 November 1677. The records of the Court of Common Council are patchy for this period and there does not seem to be any further description of Daintree's proposals.

17 London Metropolitan Archives, COL/CHD/CM/07/024, order dated 8 April 1679.

18 London Metropolitan Archives, COL/CA/01/01/084, f. 122r, order arising from a meeting held on 9 March 1675.

19 Hooke, *Posthumous Works*, p. 18.

20 Robert Hooke, *Micrographia* (London, 1665), sig. A2v.

21 Hooke, *Posthumous Works*, p. 280. For Hooke's use of architectural metaphors see Matthew C. Hunter, *Wicked Intelligence: Visual Art and the Science of Experiment in Restoration London* (Chicago, IL, and London, 2013) and Cooper, *Robert Hooke and the Rebuilding of London*, p. 123.

22 Thomas Sprat, *The History of the Royal Society of London for the Improving of Natural Knowledge* (London, 1667), p. 87.
23 Hooke, *Posthumous Works*, p. 359.
24 For a full discussion of the Cutlerian lectures see Michael Hunter, *Establishing the New Science: The Experience of the Early Royal Society* (Woodbridge, Suffolk, 1989), pp. 279–338.

3 Much Love and Service to All My Friends

1 Robert Hooke, *The Diary of Robert Hooke*, ed. Henry W. Robinson and Walter Adams (London, 1935), p. 265 (31 December 1676).
2 See, for example, Lisa Jardine, 'Robert Hooke: A Reputation Restored', in *Robert Hooke: Tercentennial Studies*, ed. Michael Cooper and Michael Hunter (Aldershot, 2006), pp. 247–58; Francesco G. Sacco, *Real, Mechanical, Experimental: Robert Hooke's Natural Philosophy* (Cham, 2020), pp. 173–5.
3 Robert Hooke, *The Posthumous Works of Robert Hooke*, ed. Richard Waller (London, 1705), p. 28.
4 London Metropolitan Archives, CLC/495/MS01758, f. 7r; this passage was omitted from the Robinson and Adams edition of Hooke's diary.
5 Hooke, *Diary*, p. 169 (12 July 1675).
6 Ibid., pp. 383 (2 November 1678), 367 (16 and 17 July 1678).
7 Ibid., pp. 205–7 (1 January 1676).
8 Ibid.
9 Roger North, *The Lives of the Right Hon. Francis North, Baron Guilford; the Hon. Sir Dudley North; and the Hon. and Rev. Dr. John North*, ed. Augustus Jessop (London, 1890), vol. I, p. 374; Thomas Birch, *The History of the Royal Society of London* (London, 1756–7), vol. III, p. 459.
10 Robert Hooke, *Early Science in Oxford*, vol. X: *The Life and Work of Robert Hooke*, ed. R. T. Gunther (Oxford, 1935), pp. 78 (3 December 1688, where 'inches' has been transcribed as 'metres'), 188 (14 February 1690), 234 (26 April 1693).
11 Hooke, *Diary*, pp. 79 (7 January 1674), 208 (3 January 1676); Hooke, *Early Science*, p. 87 (31 December 1688).
12 Quoted in Adrian Johns, *The Nature of the Book: Print and Knowledge in the Making* (Chicago, IL, and London, 1998), p. 557.
13 A. Rupert Hall and Marie Boas Hall, eds and trans., *The Correspondence of Henry Oldenburg* (Madison, WI, 1965–73, and London, 1977–86), vol. IX, p. 493.

14 This theory is stated most comprehensively in Steven Shapin, *A Social History of Truth: Civility and Science in Seventeenth-Century England* (Chicago, IL, 1994), but there are a number of persuasive arguments against it.
15 Hall and Hall, *Correspondence of Henry Oldenburg*, vol. IX, p. 493, and vol. XI, p. 178.
16 Adrian Johns, 'Flamsteed's Optics and the Identity of the Astronomical Observer', in *Flamsteed's Stars: New Perspectives on the Life and Work of the First Astronomer Royal (1646–1719)* ed. Frances Willmoth (Woodbridge, Suffolk, 1997), p. 84.
17 Hooke, *Diary*, pp. 333 (13 December 1677), 334 (14 December 1677), 336 (26 December 1677).
18 Hooke, *Posthumous Works*, p. 4.
19 Ibid., p. 11.

4 These My Poor Labours

1 Robert Hooke, *The Diary of Robert Hooke*, ed. Henry W. Robinson and Walter Adams (London, 1935), p. 369 (29 July 1678). The book was Friedrich Martens, *Spitzbergische oder Groenlandische Reise Beschreibung gethan im Jahr 1671* (Hamburg, 1675).
2 Robert Hooke, *The Posthumous Works of Robert Hooke*, ed. Richard Waller (London, 1705), p. 19.
3 Ibid., p. 18.
4 Ibid., p. 63. The original printed text has 'Will', but 'Witt' seems more plausible here.
5 Robert Hooke, *An Attempt to Prove the Motion of the Earth from Observations Made by Robert Hooke* (London, 1674), sigs. A3r–A4r.
6 Hooke, *Posthumous Works*, p. 21.
7 Robert Hooke, *Micrographia* (London, 1665), sig. g2v.
8 Christiaan Huygens to Johannes Hudde, *Oeuvres Complètes de Christiaan Huygens*, vol. V: *Correspondance 1664–1665* (La Haye, 1893), p. 304.
9 Thomas Birch, *The History of the Royal Society of London* (London, 1756–7), vol. I, pp. 219, 266, 294.
10 Hooke, *Micrographia*, p. 3.
11 Ibid., pp. 210, 211; Samuel Pepys, *The Diary of Samuel Pepys*, ed. Robert Latham and William Matthews (London, 1970–76), vol. VI, p. 18.
12 Hooke, *Posthumous Works*, p. 21; Hooke, *Micrographia*, sig. a4v.
13 Hooke, *Micrographia*, pp. 47–79. This phenomenon is now known as Newton's rings.

14 Hooke, *Attempt*, pp. 27–8.
15 Robert Hooke, 'The Substance of an Optical Discourse . . . Proposing a Way of Helping Short-Sighted or Purblind Eyes', *Philosophical Collections*, III (1681), pp. 59–60; Robert Hooke, 'The Substance of a Mechanical Discourse, Containing the Description of the Best Form of Horizontal Sails for a Mill', *Philosophical Collections*, III (1681), pp. 61–4.
16 Richard Waller, 'The Life of Dr Robert Hooke', in Hooke, *Posthumous Works*, p. xxiv.
17 Robert Knox, *An Historical Relation of the Island Ceylon, in the East Indies* (London, 1681), sig. a3r.
18 Robert Hooke, *Early Science in Oxford*, vol. X: *The Life and Work of Robert Hooke*, ed. R. T. Gunther (Oxford, 1935), p. 217 (24 February 1693).

5 A Man Who Is Mechanically Minded

1 Robert Hooke, *Micrographia* (London, 1665), pp. 142–3.
2 Ibid., p. 145.
3 Richard Waller, 'The Life of Dr Robert Hooke', in Robert Hooke, *The Posthumous Works of Robert Hooke*, ed. Richard Waller (London, 1705), p. ii.
4 John Aubrey, *Brief Lives*, ed. Kate Bennett (Oxford, 2015), vol. I, p. 97.
5 Waller, 'Life', p. iii.
6 Aubrey, *Brief Lives*, vol. I, p. 99.
7 Hooke, *Posthumous Works*, p. 39. I have emended 'severals' and 'lies' in the original printed version to read 'several' and 'lie' in the quote given here.
8 Royal Society archives, RB/1/8/30.
9 Hooke, *Posthumous Works*, p. 55.
10 Robert Hooke, *Lampas; or, Descriptions of Some Mechanical Improvements of Lamps and Waterpoises* (London, 1677), pp. 33–4. I have emended 'Reasons' in the original printed text to 'Reason' in the final sentence of this quote.
11 Hooke, *Posthumous Works*, pp. 19–20.
12 Royal Society archives, JBO/8/250. The diving ship was probably an idea rather than a fully formed design.
13 Robert Hooke, *A Description of Helioscopes, and Some Other Instruments Made by Robert Hooke* (London, 1676), p. 22.

14 Felicity Henderson, 'Unpublished Material from the Memorandum Book of Robert Hooke, Guildhall Library MS 1758', *Notes and Records of the Royal Society*, LXI (2007), pp. 129–75: pp. 135 (10 March 1672), 151 (6 October 1681); Robert Hooke, *The Diary of Robert Hooke*, ed. Henry W. Robinson and Walter Adams (London, 1935), pp. 324 (27 October 1677), 183 (29 September 1675), 182 (25 September 1675), 212 (18 January 1676).
15 Hooke, *Micrographia*, sig. e1v.
16 Ibid., sig. f2r.
17 Robert Hooke, *Lectures and Collections Made by Robert Hooke* (London, 1678), p. 99.
18 Hooke, *Diary*, p. 322 (20 October 1677).
19 Bennet Woodcroft, *Titles of Patents of Invention, Chronologically Arranged . . . 1617 . . . to . . . 1852* (London, 1854), p. 38.
20 Hooke, *Diary*, pp. 88 (23 February 1674), 282 (30 March 1677) where 'tayled' is printed for 'foyled' and 'right' for 'light'.
21 Hooke to Boyle (21 October 1664), in *The Correspondence of Robert Boyle*, vol. II: *1662–5*, ed. Michael Hunter, Antonio Clericuzio and Lawrence M. Principe (Oxford, 2001), p. 362.
22 Hooke, *Diary*, p. 100 (2 May 1674).
23 Ibid., pp. 109–10 (29 June 1674).
24 Robert Hooke, *Animadversions on the First Part of the Machina coelestis* (London, 1674), p. 54.
25 Hooke, *Diary*, pp. 148 (20 February 1675), 157 (7 April 1675), 159 (13 and 14 April 1675), 160 (9 May 1675), 203 (24 December 1675).
26 Ibid., p. 127 (printed under 20 October 1674; rightly 27 October); Hooke, *Micrographia*, p. 114.
27 See, for example, Michael Wright, 'Robert Hooke's Longitude Timekeeper' and J. A. Bennett, 'Hooke's Instruments for Astronomy and Navigation', in *Robert Hooke: New Studies*, ed. Michael Hunter and Simon Schaffer (Woodbridge, Suffolk, 1989), pp. 63–118 (p. 100) and pp. 21–32 (p. 29), respectively.
28 See Anita McConnell, 'Origins of the Marine Barometer', *Annals of Science*, LXII/1 (2005), pp. 83–101 (pp. 87–8).
29 Edmond Halley, 'An Account of Dr Robert Hook's Invention of the Marine Barometer, with Its Description and Uses, published by Order of the R[oyal] Society', *Philosophical Transactions*, XXII/269 (1701), pp. 791–4 (p. 794).
30 Robert Hooke, *An Attempt to Prove the Motion of the Earth from Observations Made by Robert Hooke* (London, 1674), sig. A4v. The historian of

science Jim Bennett has made this point: see in particular Bennett, 'Hooke's Instruments', pp. 31–2.

6 Curiosity and Beauty

1 Robert Hooke, *Micrographia* (London, 1665), pp. 153–4.
2 See Svetlana Alpers, *The Art of Describing: Dutch Art in the Seventeenth Century* (London, 1983), p. 84; Janice Neri, *The Insect and the Image: Visualizing Nature in Early Modern Europe, 1500–1700* (Minneapolis, MN, 2011), p. 113.
3 Hooke, *Micrographia*, pp. 140–41.
4 Robert Hooke, *The Posthumous Works of Robert Hooke*, ed. Richard Waller (London, 1705), p. 20.
5 Evelyn to Boyle (9 August 1659), in *The Correspondence of Robert Boyle*, vol. I: *1636–61*, ed. Michael Hunter, Antonio Clericuzio and Lawrence M. Principe (Oxford, 2001), p. 363.
6 Hooke, *Posthumous Works*, pp. 27, 24.
7 Ibid., p. 24.
8 Ibid., p. 27.
9 Ibid., pp. 27, 58, 57.
10 Ibid., p. 57.
11 Robert Hooke, *The Diary of Robert Hooke*, ed. Henry W. Robinson and Walter Adams (London, 1935), pp. 174 (14 August 1675), 176 (21 August 1675), 248 (28 August 1676), 249 (12 September 1676), 134 (9 December 1674).
12 London Metropolitan Archives CLC/495/MS01758, fol. 7v (this note was omitted from the edition of Hooke's *Diary* by Robinson and Adams).
13 Hooke, *Diary*, pp. 403 (13 and 14 March 1679), 95 (3 April 1674), 258 (19 November 1676); Robert Hooke, *Early Science in Oxford*, vol. X: *The Life and Work of Robert Hooke*, ed. R. T. Gunther (Oxford, 1935), p. 169 (3 December 1689).
14 Hooke, *Posthumous Works*, p. 26.
15 Royal Society archives CLP/20/96, fol. 1r.
16 Hooke, *Early Science*, p. 194 (6 December 1692); Hooke, *Diary*, pp. 373 (23 August 1678), 91 (11 March 1674), 440 (3 March 1680).
17 Thomas Birch, *The History of the Royal Society of London* (London, 1756–7), vol. IV, pp. 21, 25.
18 Bennet Woodcroft, *Titles of Patents of Invention, Chronologically Arranged . . . 1617 . . . to . . . 1852* (London, 1854), p. 34.

19 Hooke, *Diary*, pp. 61 (20 September 1673), 87 (17 February 1674).
20 Birch, *History*, vol. III, p. 192.
21 Hooke, *Micrographia*, p. 7.
22 Ibid., pp. 8–10; discussed in Felicity Henderson, 'Robert Hooke and the Visual World of the Early Royal Society', *Perspectives on Science*, XXVII/3 (2019), pp. 395–434.
23 John Houghton, *A Proposal for Improvement of Husbandry and Trade* (London, 1691), p. 1.

7 An Excellent System of Nature

1 Robert Hooke, *The Posthumous Works of Robert Hooke*, ed. Richard Waller (London, 1705), p. 34.
2 John Aubrey, *Brief Lives*, ed. Kate Bennett (Oxford, 2015), vol. I, p. 99.
3 Hooke, *Posthumous Works*, p. 140.
4 Ibid., p. 145.
5 Thomas Birch, *The History of the Royal Society of London* (London, 1756–7), vol. IV, p. 154.
6 Hooke, *Posthumous Works*, pp. 144–5.
7 Felicity Henderson, 'Material Thoughts: Robert Hooke's Theory of Memory', in *Testimonies: States of Mind and States of the Body in the Early Modern Period*, ed. Gideon Manning (Cham, 2020), pp. 59–83.
8 Hooke, *Posthumous Works*, pp. 64, 34.
9 Robert Hooke, *Micrographia* (London, 1665), p. 15.
10 Mark E. Erlich provides a more detailed explanation of Hooke's theory of congruity, and a useful commentary, in 'Mechanism and Activity in the Scientific Revolution: The Case of Robert Hooke', *Annals of Science*, LII/2 (1995), pp. 127–51.
11 Francesco G. Sacco, *Real, Mechanical, Experimental: Robert Hooke's Natural Philosophy* (Cham, 2020), pp. 177–8.
12 Robert Hooke, *Lectures de potentia restitutiva; or, Of Spring, Explaining the Power of Springing Bodies* (London, 1678), p. 7.
13 Hooke, *Posthumous Works*, p. 185, quoted in Sacco, *Hooke's Natural Philosophy*, p. 140. Sacco provides a much more comprehensive description of Hooke's theories of gravity and congruity than is possible here.
14 Hooke, *Posthumous Works*, p. 165.
15 Robert Hooke, *An Attempt to Prove the Motion of the Earth from Observations Made by Robert Hooke* (London, 1674), p. 27.

16 Sacco, *Hooke's Natural Philosophy*, p. 179.
17 H. W. Turnbull, ed., *The Correspondence of Isaac Newton* (Cambridge, 1960), vol. II, p. 443.
18 Sacco, *Hooke's Natural Philosophy*, p. 174.
19 Niccolò Guicciardini, 'On the Invisibility and Impact of Robert Hooke's Theory of Gravitation', *Open Philosophy*, III/1 (2020), pp. 266–82, drawing on work by Patri Pugliese and Michael Nauenberg.
20 Turnbull, ed., *Correspondence of Newton*, vol. II, p. 442, quoted in Sacco, *Hooke's Natural Philosophy*, p. 177.

8 A Discourse of Earthquakes

1 Robert Hooke, *The Posthumous Works of Robert Hooke*, ed. Richard Waller (London, 1705), p. 313.
2 Richard Waller, 'The Life of Dr Robert Hooke', ibid., p. xxiv.
3 For an excellent discussion of these themes in seventeenth-century England see William Poole, *The World Makers: Scientists of the Restoration and the Search for the Origins of the Earth* (Oxford, 2010).
4 Hooke, *Posthumous Works*, pp. 343, 291.
5 Ibid., pp. 20, 7.
6 Ibid., pp. 7–9.
7 Ibid., pp. 9–11.
8 Ibid., p. 11.
9 Thomas Browne, *Pseudodoxia epidemica*, ed. Robin Robbins, 2 vols (Oxford, 1981), vol. I, p. 1.
10 Waller, 'Life', p. xxvii.
11 Robert Hooke, *Micrographia* (London, 1665), p. 189.
12 Royal Society archives, JBO/8/236.
13 Poole, *World Makers*, p. 108.
14 Royal Society archives, JBO/9/131.
15 Royal Society archives, JBO/8/283.
16 Robert Hooke, *Early Science in Oxford*, vol. X: *The Life and Work of Robert Hooke*, ed. R. T. Gunther (Oxford, 1935), p. 263 (31 July 1693).
17 Robert Hooke, 'Some Observations, and Conjectures Concerning the Chinese Characters', *Philosophical Transactions*, XVI/180 (1686), pp. 63–78 (p. 63).
18 William Poole, 'Heterodoxy and Sinology: Isaac Vossius, Robert Hooke, and the Early Royal Society's Use of Sinology', in

The Intellectual Consequences of Religious Heterodoxy, c. 1600–1750, ed. Sarah Mortimer and John Robertson (Leiden and Boston, MA, 2012), pp. 135–53.

19 London Metropolitan Archives, CLC/495/MS01757, f. 98v.

20 Ibid.

21 Jim Bennett, 'Hooke's Instruments', in Jim Bennett, Michael Cooper, Michael Hunter and Lisa Jardine, *London's Leonardo: The Life and Work of Robert Hooke* (Oxford, 2003), pp. 73–7.

22 See A. R. Hall, 'Two Unpublished Lectures of Robert Hooke', *Isis*, XLII/3 (1951), pp. 219–30 (p. 226).

23 Hooke, *Posthumous Works*, p. 289.

24 Ibid., p. 435.

25 Ibid., p. 291.

26 Ibid., p. 323.

27 Ibid., pp. 181, 322.

28 Ibid., pp. 294–6.

29 Ibid., p. 290.

30 Ibid., pp. 294–6.

Epilogue: The Teeth of Time

1 Robert Hooke, *The Posthumous Works of Robert Hooke*, ed. Richard Waller (London, 1705), p. 433.

2 Richard Waller, 'The Life of Dr Robert Hooke', ibid., p. xxv.

3 Royal Society archives, MS/562 (3 June 1714).

4 For example, Hooke, *Posthumous Works*, pp. 83–4.

5 Robert Hooke, *Micrographia* (London, 1665), p. 210.

6 Royal Society archives, MS/847 (the 'Hooke folio'), p. 636.

SELECT BIBLIOGRAPHY

Bennett, Jim, Michael Cooper, Michael Hunter and Lisa Jardine, *London's Leonardo: The Life and Work of Robert Hooke* (Oxford, 2003)

Cooper, Michael, *Robert Hooke and the Rebuilding of London* (Stroud, 2003)

—, and Michael Hunter, eds, *Robert Hooke: Tercentennial Studies* (Aldershot, 2006)

'Espinasse, Margaret, *Robert Hooke* (London, 1956)

Guicciardini, Niccolò, *Isaac Newton and Natural Philosophy* (London, 2018)

Henderson, Felicity, 'Robert Hooke and the Visual World of the Early Royal Society', *Perspectives on Science*, XXVII/3 (2019), pp. 1–40

—, 'Material Thoughts: Robert Hooke's Theory of Memory', in *Testimonies: States of Mind and States of the Body in the Early Modern Period*, ed. Gideon Manning (Cham, 2020)

Hunter, Matthew C., *Wicked Intelligence: Visual Art and the Science of Experiment in Restoration London* (Chicago, IL, and London, 2013)

Hunter, Michael, *Establishing the New Science: The Experience of the Early Royal Society* (Woodbridge, Suffolk, 1989)

—, *Boyle: Between God and Science* (New Haven, CT, and London, 2009)

—, and Simon Schaffer, eds, *Robert Hooke: New Studies* (Woodbridge, Suffolk, 1989)

Inwood, Stephen, *The Man Who Knew Too Much: The Strange and Inventive Life of Robert Hooke, 1635–1703* (London, 2002)

Jardine, Lisa, *The Curious Life of Robert Hooke: The Man Who Measured London* (London, 2004)

Poole, William, *The World Makers: Scientists of the Restoration and the Search for the Origins of the Earth* (Oxford, 2010)

Sacco, Francesco G., *Real, Mechanical, Experimental: Robert Hooke's Natural Philosophy* (Cham, 2020)

ACKNOWLEDGEMENTS

Much of the material in this biography has arisen during work on various academic projects. Editing Hooke's diaries for Oxford University Press (forthcoming) has given me the opportunity to get to know Hooke and his day-to-day activities in a detailed way. I owe a great debt to William Poole for many Hooke-related conversations and his assistance on too many editorial points to count. I would like to thank the staff in the Royal Society Library, London, particularly Keith Moore and Rupert Baker for their long-standing friendship and research support, and Katherine Marshall for kindly allowing me to use Royal Society images in this book. Some of the material in Chapter Seven was initially presented at a conference convened by Lisa Jardine at the California Institute of Technology in Pasadena, and later published in a volume edited by Gideon Manning, *Testimonies: States of Mind and States of the Body in the Early Modern Period* (Cham, 2020). I would like to acknowledge Lisa's kind and generous support for my work on Hooke, and thank Gideon for his insightful editorial suggestions. Some of the ideas in Chapter Six were first published in a special edition of *Perspectives on Science* arising from the Arts and Humanities Research Council-funded project 'Making Visible: The Visual and Graphic Practices of the early Royal Society'. I thank my brilliant collaborators on the 'Making Visible' project, Sachiko Kusukawa, Alexander Marr, Sietske Fransen and Katherine Reinhart, for their comments and suggestions on the article in particular, but also for making the project so intellectually engaging and enjoyable. Thank you to everyone who read drafts or advised on specific questions, I'm grateful for your helpful comments. Finally, thanks and love to all my family and friends who have frequently asked how the book was going: I really appreciate your support, and I hope you like it.

PHOTO ACKNOWLEDGEMENTS

The author and publishers wish to express their thanks to the sources listed below for illustrative material and/or permission to reproduce it:

From Robert Boyle, *New Experiments Physico-Mechanicall, Touching the Spring of the Air, and Its Effects* (Oxford, 1660), photo courtesy Science History Institute, Philadelphia, pa: 2; Folger Shakespeare Library, Washington, DC (CC BY-SA 4.0): 28; from Johannes Hevelius, *Machina coelestis pars prior* (Gdańsk, 1673), photo Library of Congress, Rare Book and Special Collections Division, Washington, DC: 16; from Robert Hooke, *An Attempt to Prove the Motion of the Earth from Observations* (London, 1674), photo Smithsonian Libraries, Washington, DC: 15; from Robert Hooke, *Lectures and Collections*... (London, 1678), photo Beinecke Rare Book and Manuscript Library, Yale University, New Haven, CT: 8; The J. Paul Getty Museum, Los Angeles, CA: 21; © London Metropolitan Archives (CLC/495/MS01758): 19; from Friedrich Martens, *Spitzbergische oder Groenlandische Reise Beschreibung gethan im Jahr 1671* (Hamburg, 1675), photo ETH-Bibliothek Zurich: 11; from *Philosophical Transactions of the Royal Society*: 5 (XII/136, 25 June 1677), 27 (XX/246, November 1698); © The Royal Society: 3 (EL/PI/40), 10 (P/0010), 22 (CLP/20/40), 23 (CLP/20/96), 26 (MS/828); © The Trustees of the British Museum: 9; Wellcome Collection, London (CC by 4.0): 1, 4, 6, 7, 12, 13, 14, 17, 18, 20, 24, 25.

INDEX

Illustration numbers are indicated by *italics*